SELECTIONS FROM

Darwin's
The Origin of Species

THE SHAPE OF THE ARGUMENT

SELECTIONS FROM DARWIN'S

The Origin of Species

THE SHAPE OF THE ARGUMENT

A Science Classics Module
for Humanities Studies

Edited and annotated by Nicholas Maistrellis

Green Cat Books
an imprint of Green Lion Press
Santa Fe New Mexico

Manufactured in the United States of America

Published by Green Lion Press, Santa Fe, New Mexico

www.greenlion.com

Set in 11-point Times New Roman.
This book is printed on acid-free paper.
Printed and bound by Sheridan Books, Inc., Chelsea, Michigan.

Cover illustration: Darwin in 1881, from a photograph.
This etching appeared as the frontispiece to *The Life and Letters of Charles Darwin*, ed. Francis Darwin, vol. II (1887).

Cataloguing-in-Publication data:

Maistrellis, Nicholas
Selections from Darwin's *The Origin of Species*: The Shape of the Argument. A Science Classics Module for Humanities Studies / by Nicholas Mastrellis

Includes abridged text of Charles Darwin's *The Origin of Species*, index, introductions, biographical sketches, and notes.

ISBN 978-1888009-34-7 (softcover binding)

1. Darwin, The Origin of Species. 2. History of Science. 3. Evolution. 4. Biology.

I. Charles Darwin (1809–1882). II. Maistrellis, Nicholas (1940–). III. Title

Library of Congress Control Number: 2009922590

CONTENTS

The Green Lion's Preface

The Green Lion is pleased to offer another book in its Green Cat series of science classics for humanities studies. You can read about this series in the back of this volume on page 114.

We have long wanted to include a volume of Darwin's writings in this series. Our approach in the Green Cat books is to provide substantial passages from classic scientific texts that show the fullness and continuity of the author's argument. Too often these goals are sacrificed in anthologies. If real science is to be presented and experienced, it is not enough just to sample the flavor of an author in disconnected highlights—a continuous, coherent narrative is required.

The Origin of Species, Darwin's first book, does not present "evolution" as we know the theory today. It is the first comprehensive statement of a plausible process that would result in evolution. That process is "natural selection," which Darwin spells out in a beautifully shaped argument, with exquisite detail. *The Origin* is also of primary importance because it was so enormously influential. It continues to stand as the book you go to first—"the rest is commentary."

The substantial passages provided in this edition allow one to enjoy the precision and clarity of Darwin's writing and the patient thoroughness of his reasoning. Step by step, layer by layer, detail by detail, like the most intricate detective story, he constructs his case.

The Green Lion commissioned Nicholas Maistrellis to make the selections and provide introductions and annotations. Maistrellis knows *The Origin* deeply, has read it with students for decades, has a background in biology, and, most importantly, has studied and taught the book in the context of a broad liberal arts curriculum.

This Green Cat edition is based on the first edition of *The Origin of Species*, published in 1859. Footnotes are the editor's. We have included a thorough index, a biography of Darwin, and a bibliography and suggestions for further reading. We thank Associate Editor Howard J. Fisher for shepherding this book through and preparing the index.

Recent years have seen renewed debate concerning natural selection, evolution, and other aspects of Darwin's thought. We hope this volume will contribute to the conversation by revealing the real foundations of the argument and showing the thoroughness with which Darwin develops his case. Darwin's achievement is not the formulation of an idea; rather, it is the assembling of a wide variety of ordinary and unobjectionable facts into a powerful statement of what *must necessarily* follow from these facts. This is not just an idea, and it can't be fruitfully discussed as just an idea. In addition, it is both interesting and instructive to see that Darwin is far from doctrinaire in his presentation. He leaves the discussion as a continuing one. It continues today, and today's readers and students can be part of it.

Dana Densmore
William H. Donahue
February 2009

About Charles Darwin

In his landmark work, *The Origin of Species*, Charles Robert Darwin (1809–1882) argued that all living beings have descended from common ancestors through the process called natural selection. This came to be called the Theory of Evolution, although Darwin never used the term "evolution" in *The Origin*.

Darwin's early life did not suggest the eminence he would later achieve. He was sent to study medicine at the University of Edinburgh, but his sensibilities were so offended by the horrors of the mid-nineteenth century operating room in the absence of anesthesia that a medical career became impossible for him. Then as a last resort he was sent to Cambridge to study divinity. Here he exhibited a lack of interest in theology but developed a growing passion for natural history, especially the study of beetles, in which he became quite expert.

In 1831 he was invited to join the naval vessel HMS Beagle as naturalist and companion to Capt. Robert Fitzroy. The Beagle was commissioned to survey the coast of South America. This was a position Darwin greatly desired, though it was far from what his family had planned for him. His father's objections were overcome through the intercession of Darwin's uncle Josiah Wedgwood, who argued that "the pursuit of Natural History, though certainly not professional, is very suitable to a Clergyman." After being interviewed by Fitzroy and accepted for the voyage, Darwin is said to have cried out, "Woe unto ye beetles of South America!"

Soon after his return to England in 1836 Darwin began the work that culminated in the publication of *The Origin of Species* in November 1859. The book was an instant sensation. The first printing of 1,250 copies was sold out in advance, and a new printing was quickly undertaken. Almost two hundred reviews appeared within the first few months of its appearance. *The Origin* attracted both respectful attention and also fierce opposition. In spite of Darwin's caution about mentioning the history of man, the implications of the theory for the understanding of human origins were discerned very quickly, especially by those who believed that any theory of animal origins for human nature undermined morality.

There was also opposition from within the scientific community. Although most scientists agreed that Darwin had made an overwhelming case for evolution as against special creation, many were unconvinced by his theory of natural selection. A number of objections were raised to Darwin's premises. Many naturalists were not persuaded that gaps in the fossil record were due to incompleteness, but thought they were evidence of genuine discontinuities—evidence that new species had emerged suddenly, rather than gradually as Darwin had proposed. The eminent physicist Lord Kelvin presented evidence from mathematical physics that the earth could not be as old as Darwin thought necessary for natural selection to work.

In the late nineteenth century many paleontologists began to favor interpretation of the fossil record as a series of straight-line developments caused by factors internal to the organism, rather than by natural selection. There continued to be little support for natural selection in the early decades of the 20th century. This changed only when a number of scientists in the 1920's and 1930's, among them R. A. Fisher, J. B. S. Haldane, and Sewall Wright, showed how to combine Darwin's insights with the newly-developing science of genetics to produce a theory of natural selection that both met the current objections and also opened new directions of research. This work culminated in what became known as the "Modern Synthesis" of genetics and evolutionary theory in the late 1930's.

Darwin undertook no further voyages of discovery. He spent the rest of his life with his wife and family in their country home in Down, Kent, continuing the study of living things, conducting experiments in his garden, and writing a remarkable series of major works on an extraordinary variety of topics. These include *The Descent of Man, and Selection in Relation to Sex* (1871), in which Darwin finally marshaled the evidence for the origin of man from other animals and also presented evidence for another form of selection, "sexual selection," the selective force exerted by female preference for males of certain types. This book essentially completed his theoretical work on evolution. He also produced a series of investigations on mechanisms of self- and cross-fertilization in plants, culminating in three books, *The Effects of Cross and Self Fertilization in the Vegetable Kingdom* (1876), *The Different Forms of Flowers on Plants of the Same Species* (1877), and *The Various Contrivances by Which Orchids are Fertilized by Insects* (1877). His final book, published the year before his death, was *The Formation of Vegetable Mold Through the Action of Worms, With Observations on their Habits* (1881). In this book Darwin returned to an old theme, the production of large effects through the accumulation of small causes—in this case, the power of worms to bury whole abandoned cities through their digging.

The year 2009 is both the bicentenary of Darwin's birth and the 150th anniversary of the publication of *The Origin*. Darwin shares his birth date of February 12, 1809 with Abraham Lincoln.

Editor's General Introduction

Darwin's *Origin of Species*, one of the great foundational works of modern science, reorganized biology according to a revolutionary idea—that the world of living things that surrounds us, and of which we are a part, is essentially *historical*; that it has come into being through time, and by means of causes that act in time. Such a statement may seem obscure at first; but the vision it represents will become increasingly evident as we follow Darwin's discussion.

Although it is a scientific work, the *Origin* was not written solely for specialists. Darwin expected it to be read and understood by generally educated readers, readers like you and me. In Chapter XIV he will state that "this whole volume is one long argument." For the present edition I have chosen selections that attempt to present Darwin's principal lines of argument, while of course passing over many details. I have also provided notes and other remarks designed to help you focus on what is essential in Darwin's argument for his theory of the development of living things.

In the complete *Origin of Species*, Darwin exhibited his theory in the first five chapters. In the next eight chapters he described both the difficulties and the power of the theory; the final chapter summarized the whole and looked ahead to the transformation, under his theory, of the science of life. Our selections preserve that organization, concentrating first on the laying out of the theory, then focusing on two moments from Chapters VI and XI which illustrate the challenges Darwin thought the theory must face, as well as its power to explain important facts about the living world.

As you begin your reading of the *Origin*, think carefully about the two epigraphs which Darwin placed opposite the title page of the first edition:

> But with regard to the material world, we can at least go so far as this—we can perceive that events are brought about not by insulated interpositions of Divine powers exerted in each particular case, but by the establishment of general laws.
>
> W. WHEWELL: *Bridgewater Treatise*

> To conclude, therefore, let no man out of a weak conceit of sobriety, or an ill-applied moderation, think or maintain, that a man can search too far or be too well studied in the book of God's word, or in the book of God's works—divinity or philosophy—but rather let men endeavour an endless progress or proficience in both.
>
> BACON: *Advancement of Learning*

What do these statements mean, and why did Darwin choose them? The passage by Whewell, who was Darwin's contemporary and a well-known philosopher, claims that the world was divinely created, but that it can be understood only by the scientist, not the theologian—for God so made the world that the miracle of its creation is in the background, whereas what is presented to our senses and understanding is something whose structure can be discerned. The passage by Francis Bacon is somewhat different in substance, although similar in tone. Bacon, who lived more than 200 years before Darwin and is one of the founders of modern science, argues that both natural science and biblical study are worthy of the most intense human efforts. Neither of the two quotations speaks directly about the origin of species. They seem rather to be efforts to assure us both of the author's religiosity and of the propriety, from a religious point of view, of devoting oneself to scientific investigation.

Pay careful attention also to Darwin's title. Here is the way he presented it on the title page of the first edition:

THE ORIGIN OF SPECIES

BY MEANS OF NATURAL SELECTION

OR THE

PRESERVATION OF FAVOURED RACES IN THE STRUGGLE FOR LIFE

Does the "OR" between title and subtitle mean that the latter is a restatement of the former? The words *species* and *race* may need some explanation. For Darwin, species were just the kinds of living things that we see around us—not "trout" generally, for example, but "rainbow trout" and "brown trout" are species. In the same way, "white oak" and "red oak" are species; "oak" is not. Nevertheless, "trout" and "oak" are not insignificant words. In ordinary speech these groupings, called *genera* by scientists (singular, *genus*), denote groups of species of animals and plants that are so much alike that they seem to constitute a group apart from others, for example, the trout as opposed to the bass and the perch, or the oaks as opposed to the maples and pines. But for Darwin, these groupings signified more than resemblance. They implied community of descent, that is, having a common ancestor. For Darwin, community of descent became the crucial fact about living beings, while resemblance was only an indicator of common origin. Moreover it was community of descent, Darwin believed, that explained the subordination of groups under larger groups. He represented this hierarchy through the image of the *Tree of Life*, in which groups were compared to branches originating from a common stem. This image will come up again and again in the book, and it should be closely attended to whenever it does.

As for race, it is almost impossible in our time to use the word in a way that is not politically charged. But in Darwin's day it had a noncontroversial use; it simply referred to local populations that were members of the same species yet sufficiently different from one another, and sufficiently coherent as groups, to deserve notice and a distinctive name. Sometimes Darwin refers to "geographical races," indicating that they are tied to a particular locale. For example, the whitetail deer is a species of deer widely distributed throughout North America. However, the deer populations in different parts of their range differ enough from one another that naturalists and hunters who know them well can often tell from what part of the country a particular deer came—that is, to what geographical race it belongs. Darwin will argue that a new species arises from a local population of an old species that has been favored in some way. How that happens is the principal topic of the *Origin*. Perhaps it has occurred to you that the relation between "race" and "species" is, formally, very much like that between "species" and "genus." It is indeed Darwin's radical contention, as indicated in the title, that the former turns into the latter.

In his Introduction Darwin begins the inquiry by distinguishing between a population becoming *adapted to a way of life*, and different populations becoming *co-adapted to one another*. This distinction is crucial. If, for example, you see an animal growing a longer and thicker fur coat as it moves into colder climates, and its offspring doing the same, then it is possible to argue that the colder environment is that to which the animal has become adapted—that the environment has caused the change. But when you see, on the one hand, a plant whose flowers have separate sexes and which need an insect to fertilize them, and on the other hand an insect whose body has to have a particular shape in order to enter the flowers of that plant to get food, the question of causality becomes more difficult. There is no longer a stable environment that can be identified as the cause of these facts; rather we seem to have two living things which have developed in relation to one another. The idea of co-adaptation opens up to us a picture of the world in which the mutual relations of living things are not fixed, but are capable of change.

INTRODUCTION

WHEN on board H.M.S. Beagle, as naturalist, I was much struck with certain facts in the distribution of the inhabitants of South America, and in the geological relations of the present to the past inhabitants of that continent. These facts seemed to me to throw some light on the origin of species—that mystery of mysteries, as it has been called by one of our greatest philosophers.[1]

* * *

In considering the Origin of Species, it is quite conceivable that a naturalist, reflecting on the mutual affinities of organic beings, on their embryological relations, their geographical distribution, geological succession, and other such facts, might come to the conclusion that each species had not been independently created, but had descended, like varieties, from other species. Nevertheless, such a conclusion, even if well founded, would be unsatisfactory, until it could be shown how the innumerable species inhabiting this world have been modified, so as to acquire that perfection of structure and coadaptation which most justly excites our admiration. Naturalists continually refer to external conditions, such as climate, food, &c., as the only possible cause of variation. In one very limited sense, as we shall hereafter see, this may be true; but it is preposterous to attribute to mere external conditions, the structure, for instance, of the woodpecker, with its feet, tail, beak, and tongue, so admirably adapted to catch insects under the bark of trees. In the case of the misseltoe, which draws its nourishment from certain trees, which has seeds that must be transported by certain birds, and which has flowers with separate sexes absolutely requiring the agency of certain insects to bring pollen from one flower to the other, it is equally preposterous to account for the structure of this parasite, with its relations to several distinct organic beings, by the effects of external conditions, or of habit, or of the volition of the plant itself.

1. *One of our greatest philosophers*: Darwin refers to John Herschel (1792–1871), English mathematician, astronomer, and chemist. He had written, "Of course I allude to that mystery of mysteries, the replacement of extinct species by others," in a letter that was published in 1838.

The author of the *Vestiges of Creation*[2] would, I presume, say that, after a certain unknown number of generations, some bird had given birth to a woodpecker, and some plant to the misseltoe, and that these had been produced perfect as we now see them; but this assumption seems to me to be no explanation, for it leaves the case of the coadaptations of organic beings to each other and to their physical conditions of life, untouched and unexplained.

It is, therefore, of the highest importance to gain a clear insight into the means of modification and coadaptation. At the commencement of my observations it seemed to me probable that a careful study of domesticated animals and of cultivated plants would offer the best chance of making out this obscure problem. Nor have I been disappointed; in this and in all other perplexing cases I have invariably found that our knowledge, imperfect though it be, of variation under domestication, afforded the best and safest clue. I may venture to express my conviction of the high value of such studies, although they have been very commonly neglected by naturalists.

From these considerations, I shall devote the first chapter of this Abstract to Variation under Domestication. We shall thus see that a large amount of hereditary modification is at least possible; and, what is equally or more important, we shall see how great is the power of man in accumulating by his Selection successive slight variations. I will then pass on to the variability of species in a state of nature; but I shall, unfortunately, be compelled to treat this subject far too briefly, as it can be treated properly only by giving long catalogues of facts. We shall, however, be enabled to discuss what circumstances are most favourable to variation. In the next chapter the Struggle for Existence amongst all organic beings throughout the world, which inevitably follows from their high geometrical powers of increase, will be treated of. This is the doctrine of Malthus, applied to the whole animal and vegetable kingdoms. As many more individuals of each species are born than can possibly survive; and as, consequently, there is a frequently recurring struggle for existence, it follows that any being, if it vary however slightly in any manner profitable to itself, under the complex and sometimes varying conditions of life, will have a better chance of surviving, and thus be naturally *selected.* From the strong principle of inheritance, any selected variety will tend to propagate its new and modified form.

2. "The author of the *Vestiges of Creation*" was Robert Chambers, who in that book proposed a theory of the development of higher life forms from lower—without, however, specifying by what process he supposed such development to have taken place. Darwin, writing in 1859, does not identify him by name because *Vestiges* had been published anonymously (in 1844). The fact of Chambers' authorship remained concealed until 1884, some thirteen years after Chambers' death.

This fundamental subject of Natural Selection will be treated at some length in the fourth chapter; and we shall then see how Natural Selection almost inevitably causes much Extinction of the less improved forms of life, and induces what I have called Divergence of Character. In the next chapter I shall discuss the complex and little known laws of variation and of correlation of growth. In the four succeeding chapters, the most apparent and gravest difficulties on the theory will be given: namely, first, the difficulties of transitions, or in understanding how a simple being or a simple organ can be changed and perfected into a highly developed being or elaborately constructed organ; secondly, the subject of Instinct, or the mental powers of animals; thirdly, Hybridism, or the infertility of species and the fertility of varieties when intercrossed; and fourthly, the imperfection of the Geological Record. In the next chapter I shall consider the geological succession of organic beings throughout time; in the eleventh and twelfth, their geographical distribution throughout space; in the thirteenth, their classification or mutual affinities, both when mature and in an embryonic condition. In the last chapter I shall give a brief recapitulation of the whole work, and a few concluding remarks.

No one ought to feel surprise at much remaining as yet unexplained in regard to the origin of species and varieties, if he makes due allowance for our profound ignorance in regard to the mutual relations of all the beings which live around us. Who can explain why one species ranges widely and is very numerous, and why another allied species has a narrow range and is rare? Yet these relations are of the highest importance, for they determine the present welfare, and, as I believe, the future success and modification of every inhabitant of this world. Still less do we know of the mutual relations of the innumerable inhabitants of the world during the many past geological epochs in its history. Although much remains obscure, and will long remain obscure, I can entertain no doubt, after the most deliberate study and dispassionate judgement of which I am capable, that the view which most naturalists entertain, and which I formerly entertained—namely, that each species has been independently created—is erroneous. I am fully convinced that species are not immutable; but that those belonging to what are called the same genera are lineal descendants of some other and generally extinct species, in the same manner as the acknowledged varieties of any one species are the descendants of that species. Furthermore, I am convinced that Natural Selection has been the main but not exclusive means of modification.

Editor's Introduction to Chapter I

The idea of selection.

Darwin was unusual in his time, even among those who were actively thinking about the origin of species, in believing that the work of practical gardeners and animal breeders could be of interest to the theoretical biologist. Indeed, at first sight there does not seem to be much connection between the modification of plants and animals to suit human needs and tastes and the natural transformations of wild plants and animals over historical time; there is, after all, no "breeder" in nature. Nevertheless, not only did Darwin think that gardeners and breeders could provide valuable information for theorists, he structured his entire argument as a long and complex analogy between the works of domestication and the works of nature. The basis of that analogy is the idea of *selection*.

Gardeners and stockmen regularly take note of plants and animals that they find valuable or amusing, and they breed exclusively from them. Thus, domestic breeding involves two activities: first, particular organisms are noticed, prized and specially attended to; second, only such chosen, or "selected," specimens are allowed to reproduce. It was Darwin's main contention that something similar must be taking place in the natural world, even though there is no human breeder to be found there.

Varieties, sub-varieties, and breeds

It will be easier to follow Darwin's reasoning if we first clarify some additional terminology. The chapter begins with a question about cultivated *varieties* and *sub-varieties* of animals and plants; later Darwin will use the term *breed.* As a first approximation these terms are interchangeable, although a sub-variety is always subordinate to some variety. Darwin calls the various kinds of cultivated animals and plants "breeds" when he wants to stress that they are human productions, the results of systematic cultivation from some progenitor. For example, the mastiff is a variety or breed of dog, while the Brazilian mastiff is a sub-variety or breed of mastiff.

The word *species* is used very sparingly by Darwin when talking about cultivated animals and plants. He thinks it a great question whether human breeders have ever created a new species.

Variation

Darwin's argument in the first third of Chapter I is both subtle and difficult. He is dealing with three separate questions at once, thinking of them nevertheless as parts of a single larger issue, that of *variation*. The three questions are: (1) What is the origin or cause of variation? (2) What causes the incidence of variation within a living population to increase or decrease? and (3) What distinguishes between heritable and nonheritable variations?

A modern geneticist would be tempted to begin with the third question, regarding its answer as straightforward and, moreover, as determining the answers to the other two questions. In the modern understanding, heritable changes in an organism are those that can be traced to changes in the DNA of the reproductive cells of that organism; all other changes die with the death of the living being and are not passed on to its offspring. According to this view, then, the first two questions will also have clear answers: first, that variation is ultimately caused by replication errors, or other kinds of rearrangement, in an organism's DNA; second, that the incidence of these changes is affected in complex ways by both internal and environmental factors.

If we are to understand Darwin we must put such doctrines aside. We must especially set aside any conviction that changes acquired by a living thing in its lifetime—for example, the development of longer fur—*cannot* be inherited by its offspring; for to Darwin the question whether acquired characteristics can be inherited was still an open one. Darwin possessed no theory about inheritance like that of modern genetics; he had, therefore, to begin with the phenomena of variation themselves. Now, cultivated animals and plants seem to be more variable than are those in the wild, and their conditions of life seem also to be more variable—certainly their conditions of life differ from those of the wild animals and plants from which they originated. Could it be, he asked, that the one circumstance causes the other—the alterations in their living conditions being responsible for the greater variability of domestic productions? And in general, could changes in the conditions of life be responsible for the existence of variation among organisms?

To investigate this one must ask *how* these variations are related to the changes in living conditions. It is observed, for example, that animals subjected to colder climates grow longer fur, that birds that are prevented from flying finally become unable to fly, and so on. Are these changes typical, and are they inherited? Darwin found that bodily changes which are directly correlated with changes in the environment, or with changes in habits, *sometimes* seem to be inherited, but that these are not the typical cases. Instead he suspected that changes in habits and living conditions caused the whole reproductive system to become unstable, and therefore more likely to put forth offspring that differed from their parents in all sorts of ways not directly correlated with the particular changes in the conditions of life or habits themselves. A modern biologist would call these changes "random," meaning precisely that they seem to have no connection with the needs of the organism—that they just occur.

The important thing to notice is that Darwin has concluded that these kinds of changes are *much more common* than the directed or correlated ones (such as growing a longer coat in a colder climate, or losing the ability to fly through abandoning the habit of flying). It is precisely these interesting, unexpected, and apparently "random" changes that make selection possible and thus give breeders the power to alter the breeds of domesticated animals and plants. Without those changes domestication would be far more restricted; for in their absence it would be hard or impossible to produce an organism that is not adapted to its conditions of life and so requires constant human care in order to survive—as is indeed true of many domesticated breeds.

Pigeons

Chapter I pays special attention to the question of the origin of various breeds of pigeons; and Darwin gives his reasons for believing that the common opinion of naturalists is correct—that all existing pigeon varieties ultimately arose from a single wild progenitor, the rock dove (*Columba livia*). We may easily observe examples of the rock dove, since the pigeons commonly found in parks and cities in the United States belong to this very variety. They are not native to North America but were brought over by English settlers to raise for food, not for amusement. According to Darwin's principles, that is why they do not have the peculiar traits of the special breeds he discusses; for when these pigeons ceased being useful they were released, subsequently establishing semi-wild populations. Take advantage of any opportunities you may have to pay attention to common pigeons. Notice their habits and the ways in which they differ from one another.

Although Darwin in the *Origin* did not provide illustrations of the various pigeon breeds he discusses, I have reproduced lithographs of several of those breeds from one of Darwin's later books in order to assist you in following Darwin's descriptions. Try not to be overwhelmed by the numerous technical terms of pigeon anatomy; rather, focus on the central argument—that all breeds of domesticated pigeons descended originally from one wild species, *Columba livia*, the rock dove. The main reasons supporting this conclusion are, first, that all domesticated pigeons share many characteristics with *C. livia*—for example, the tendency to nest on cliffs and not in trees. Second, domesticated breeds when crossed with one another often put forth young which more directly resemble the wild rock dove than do their parents. As an example, the rock dove's characteristic slate blue color frequently appears in the offspring of crossed breeds that are not themselves slate blue—and it was well-known to scientists of Darwin's day that ancestral traits could be hidden in descendant organisms only to reappear in subsequent generations. A third reason is that there are no other wild pigeons that are sufficiently like the rock dove to be alternative candidates for possible ancestors.

Darwin regards it as an important cautionary observation that all breeders disagree with naturalists and believe that each of their breeds has descended from a separate wild species—so taken are they with the differences among their breeds!

CHAPTER I

VARIATION UNDER DOMESTICATION

WHEN we look to the individuals of the same variety or subvariety of our older cultivated plants and animals, one of the first points which strikes us, is, that they generally differ much more from each other, than do the individuals of any one species or variety in a state of nature. When we reflect on the vast diversity of the plants and animals which have been cultivated, and which have varied during all ages under the most different climates and treatment, I think we are driven to conclude that this greater variability is simply due to our domestic productions having been raised under conditions of life not so uniform as, and somewhat different from, those to which the parent-species have been exposed under nature. There is, also, I think, some probability in the view propounded by Andrew Knight, that this variability may be partly connected with excess of food. It seems pretty clear that organic beings must be exposed during several generations to the new conditions of life to cause any appreciable amount of variation; and that when the organisation has once begun to vary, it generally continues to vary for many generations. No case is on record of a variable being ceasing to be variable under cultivation. Our oldest cultivated plants, such as wheat, still often yield new varieties: our oldest domesticated animals are still capable of rapid improvement or modification.

It has been disputed at what period of life the causes of variability, whatever they may be, generally act; whether during the early or late period of development of the embryo, or at the instant of conception. Geoffroy St. Hilaire's experiments[3] show that unnatural treatment of the embryo causes monstrosities; and monstrosities cannot be separated by any clear line of distinction from mere variations. But I am strongly inclined to suspect that the most frequent cause of variability may be attributed to the male and female reproductive elements having been affected prior to the act of conception. Several reasons make me believe in this; but the chief one is the remarkable effect which confinement or cultivation has on

3. Étienne Geoffroy St. Hilaire (1772–1844) was a French naturalist and zoologist. One series of his experiments involved injury of chick embryos by shaking, perforating, or depriving the egg of air during incubation. "The constant effect of these injuries was the production of a very considerable number of anomalies..." (*Penny Cyclopaedia* of the Society for the Diffusion of Useful Knowledge, 1839).

the functions of the reproductive system; this system appearing to be far more susceptible than any other part of the organization, to the action of any change in the conditions of life. Nothing is more easy than to tame an animal, and few things more difficult than to get it to breed freely under confinement, even in the many cases when the male and female unite. How many animals there are which will not breed, though living long under not very close confinement in their native country! This is generally attributed to vitiated instincts; but how many cultivated plants display the utmost vigour, and yet rarely or never seed! In some few such cases it has been found out that very trifling changes, such as a little more or less water at some particular period of growth, will determine whether or not the plant sets a seed. I cannot here enter on the copious details which I have collected on this curious subject; but to show how singular the laws are which determine the reproduction of animals under confinement, I may just mention that carnivorous animals, even from the tropics, breed in this country pretty freely under confinement, with the exception of the plantigrades or bear family; whereas, carnivorous birds, with the rarest exceptions, hardly ever lay fertile eggs. Many exotic plants have pollen utterly worthless, in the same exact condition as in the most sterile hybrids. When, on the one hand, we see domesticated animals and plants, though often weak and sickly, yet breeding quite freely under confinement; and when, on the other hand, we see individuals, though taken young from a state of nature, perfectly tamed, long-lived, and healthy (of which I could give numerous instances), yet having their reproductive system so seriously affected by unperceived causes as to fail in acting, we need not be surprised at this system, when it does act under confinement, acting not quite regularly, and producing offspring not perfectly like their parents or variable.

Sterility has been said to be the bane of horticulture; but on this view we owe variability to the same cause which produces sterility; and variability is the source of all the choicest productions of the garden. I may add, that as some organisms will breed most freely under the most unnatural conditions (for instance, the rabbit and ferret kept in hutches), showing that their reproductive system has not been thus affected; so will some animals and plants withstand domestication or cultivation, and vary very slightly—perhaps hardly more than in a state of nature.

* *

Habit also has a deciding influence, as in the period of flowering with plants when transported from one climate to another. In animals it has a more marked effect; for instance, I find in the domestic duck that the bones of the wing weigh less and the bones of the leg more, in proportion to the whole skeleton, than do the same bones in the wild-duck; and I presume that this change may be safely attributed to the domestic duck flying much less, and walking more, than its wild parent. The great and inherited development of the udders in cows and goats in countries where they are habitually milked, in comparison with the state of these organs in other countries, is another instance of the effect of use. Not a single domestic animal can be named which has not in some country drooping ears; and the view suggested by some authors, that the drooping is due to the disuse of the muscles of the ear, from the animals not being much alarmed by danger, seems probable.

There are many laws regulating variation, some few of which can be dimly seen, and will be hereafter briefly mentioned. I will here only allude to what may be called correlation of growth. Any change in the embryo or larva will almost certainly entail changes in the mature animal. In monstrosities, the correlations between quite distinct parts are very curious; and many instances are given in Isidore Geoffroy St. Hilaire's[4] great work on this subject. Breeders believe that long limbs are almost always accompanied by an elongated head. Some instances of correlation are quite whimsical; thus cats with blue eyes are invariably deaf; colour and constitutional peculiarities go together, of which many remarkable cases could be given amongst animals and plants. From the facts collected by Heusinger, it appears that white sheep and pigs are differently affected from coloured individuals by certain vegetable poisons. Hairless dogs have imperfect teeth; long-haired and coarse-haired animals are apt to have, as is asserted, long or many horns; pigeons with feathered feet have skin between their outer toes; pigeons with short beaks have small feet, and those with long beaks large feet. Hence, if man goes on selecting, and thus augmenting, any peculiarity, he will almost certainly unconsciously modify other parts of the structure, owing to the mysterious laws of the correlation of growth.

The result of the various, quite unknown, or dimly seen laws of variation is infinitely complex and diversified. It is well worth while carefully to study the several treatises published on some of our old cultivated plants, as on the hyacinth, potato, even the dahlia, &c.; and it is really surprising to note the endless points in structure and constitution in which the varieties and subvarieties differ slightly from each other. The whole organization seems to have become plastic, and tends to depart in some small degree from that of the parental type.

Any variation which is not inherited is unimportant for us. But the number and diversity of inheritable deviations of structure, both those of slight

4. Isidore Geoffroy St. Hilaire (1805–1861), son of Étienne (page 13*n*), continued to develop many of his father's ideas, and both father and son were regularly cited by Darwin.

English Carrier

Short-faced English Tumbler

from *The Variation of Animals and Plants under Domestication*, 1875

and those of considerable physiological importance, is endless. Dr. Prosper Lucas's treatise,[5] in two large volumes, is the fullest and the best on this subject. No breeder doubts how strong is the tendency to inheritance: like produces like is his fundamental belief: doubts have been thrown on this principle by theoretical writers alone. When a deviation appears not unfrequently, and we see it in the father and child, we cannot tell whether it may not be due to the same original cause acting on both; but when amongst individuals, apparently exposed to the same conditions, any very rare deviation, due to some extraordinary combination of circumstances, appears in the parent—say, once amongst several million individuals—and it reappears in the child, the mere doctrine of chances almost compels us to attribute its reappearance to inheritance. Every one must have heard of cases of albinism, prickly skin, hairy bodies, &c., appearing in several members of the same family. If strange and rare deviations of structure are truly inherited, less strange and commoner deviations may be freely admitted to be inheritable. Perhaps the correct way of viewing the whole subject, would be, to look at the inheritance of every character whatever as the rule, and non-inheritance as the anomaly.

* * *

On the Breeds of the Domestic Pigeon. Believing that it is always best to study some special group, I have, after deliberation, taken up domestic pigeons. I have kept every breed which I could purchase or obtain, and have been most kindly favoured with skins from several quarters of the world, more especially by the Hon. W. Elliot from India, and by the Hon. C. Murray from Persia. Many treatises in different languages have been

5. Prosper Lucas (1805–1885) was a French medical doctor specializing in the study of heredity. The two volumes of his *Philosophical and Physiological Treatise on Natural Heredity* were published in 1847 and 1850.

English Barb English Pouter

published on pigeons, and some of them are very important, as being of considerably antiquity. I have associated with several eminent fanciers, and have been permitted to join two of the London Pigeon Clubs. The diversity of the breeds is something astonishing. Compare the English carrier and the short-faced tumbler, and see the wonderful difference in their beaks, entailing corresponding differences in their skulls. The carrier, more especially the male bird, is also remarkable from the wonderful development of the caruncculated skin about the head, and this is accompanied by greatly elongated eyelids, very large external orifices to the nostrils, and a wide gape of mouth. The short-faced tumbler has a beak in outline almost like that of a finch; and the common tumbler has the singular and strictly inherited habit of flying at a great height in a compact flock, and tumbling in the air head over heels. The runt is a bird of great size, with long, massive beak and large feet; some of the sub-breeds of runts have very long necks, others very long wings and tails, others singularly short tails. The barb is allied to the carrier, but, instead of a very long beak, has a very short and very broad one. The pouter has a much elongated body, wings, and legs; and its enormously developed crop, which it glories in inflating, may well excite astonishment and even laughter. The turbit has a very short and conical beak, with a line of reversed feathers down the breast; and it has the habit of continually expanding slightly the upper part of the œsophagus. The Jacobin has the feathers so much reversed along the back of the neck that they form a hood, and it has, proportionally to its size, much elongated wing and tail feathers. The trumpeter and laugher, as their names express, utter a very different coo from the other breeds. The fantail has thirty or even forty tail-feathers, instead of twelve or fourteen, the normal number in all members of the great pigeon family; and these feathers are kept expanded, and are carried so erect that in good birds the head and

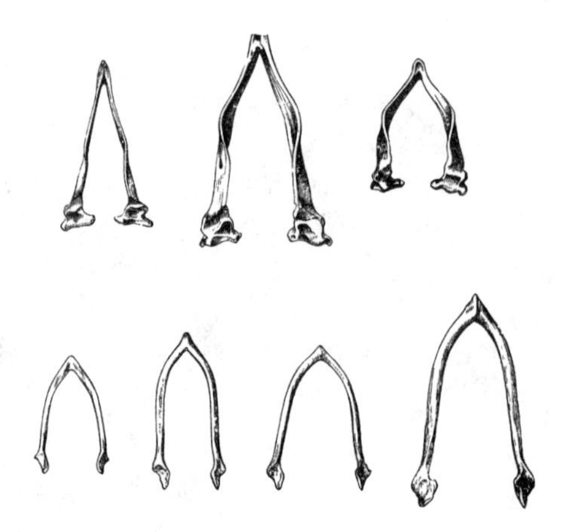

Above: The lower jaws of several pigeon breeds. *Below*: Pigeon furculae from *The Variation of Animals and Plants under Domestication*, 1875

tail touch; the oil-gland is quite aborted. Several other less distinct breeds might have been specified.

In the skeletons of the several breeds, the development of the bones of the face in length and breadth and curvature differs enormously. The shape, as well as the breadth and length of the ramus[6] of the lower jaw, varies in a highly remarkable manner. The number of the caudal and sacral vertebrae[7] vary; as does the number of the ribs, together with their relative breadth and the presence of processes. The size and shape of the apertures in the sternum[8] are highly variable; so is the degree of divergence and relative size of the two arms of the furcula.[9] The proportional width of the gape of mouth, the proportional length of the eyelids, of the orifice of the nostrils, of the tongue (not always in strict correlation with the length of beak), the size of the crop and of the upper part of the œsophagus; the development and abortion of the oil-gland; the number of the primary wing and caudal feathers; the relative length of wing and tail to each other and to the body;

6. The *ramus* is the posterior process of the lower jaw. See Darwin's sketches of pigeon jaws at the top of this page.

7. *vertebrae*: In the axial skeleton of vertebrates, the bony segments forming the spinal column or backbone. Sacral vertebrae fuse with the pelvis, allowing the transfer of force to the skeleton as a whole. Caudal vertebrae are smaller and less specialized, forming the tail of the organism. (Adapted from Michael Allaby, *A Dictionary of Zoology*, 1999.)

8. The *sternum* is the long flat bone located in the center of the chest of vertebrates. Together with the rib bones (to which it is joined by cartilage), it forms the rib cage.

9. *furcula* here refers to what is commonly called the "wishbone." More generally, it denotes any forked organ or structure. See Darwin's drawings of pigeon furculae above on this page.

English Fantail

Rock-Pigeon

the relative length of leg and of the feet; the number of scutellae[10] on the toes, the development of skin between the toes, are all points of structure which are variable. The period at which the perfect plumage is acquired varies, as does the state of the down with which the nestling birds are clothed when hatched. The shape and size of the eggs vary. The manner of flight differs remarkably; as does in some breeds the voice and disposition. Lastly, in certain breeds, the males and females have come to differ to a slight degree from each other.

Altogether at least a score of pigeons might be chosen, which if shown to an ornithologist, and he were told that they were wild birds, would certainly, I think, be ranked by him as well-defined species. Moreover, I do not believe that any ornithologist would place the English carrier, the short-faced tumbler, the runt, the barb, pouter, and fantail in the same genus; more especially as in each of these breeds several truly-inherited sub-breeds, or species as he might have called them, could be shown him.

Great as the differences are between the breeds of pigeons, I am fully convinced that the common opinion of naturalists is correct, namely, that all have descended from the rock-pigeon (*Columba livia*), including under this term several geographical races or sub-species, which differ from each other in the most trifling respects. As several of the reasons which have led me to this belief are in some degree applicable in other cases, I will here briefly give them. If the several breeds are not varieties, and have not proceeded from the rock-pigeon, they must have descended from at least seven or eight aboriginal stocks; for it is impossible to make the present domestic

10. *scutellum* (plur. *scutellae*): From Latin *scutum* (shield). Here, one of the horny scales on a bird's foot; more generally, any scale, plate, or shieldlike formation.

breeds by the crossing of any lesser number: how, for instance, could a pouter be produced by crossing two breeds unless one of the parent-stocks possessed the characteristic enormous crop? The supposed aboriginal stocks must all have been rock-pigeons, that is, not breeding or willingly perching on trees. But besides *C. livia*, with its geographical subspecies, only two or three other species of rock-pigeons are known; and these have not any of the characters of the domestic breeds. Hence the supposed aboriginal stocks must either still exist in the countries where they were originally domesticated, and yet be unknown to ornithologists; and this, considering their size, habits, and remarkable characters, seems very improbable; or they must have become extinct in the wild state. But birds breeding on precipices, and good fliers, are unlikely to be exterminated; and the common rock-pigeon, which has the same habits with the domestic breeds, has not been exterminated even on several of the smaller British islets, or on the shores of the Mediterranean. Hence the supposed extermination of so many species having similar habits with the rock-pigeon seems to me a very rash assumption. Moreover, the several above-named domesticated breeds have been transported to all parts of the world, and, therefore, some of them must have been carried back again into their native country; but not one has ever become wild or feral, though the dovecot-pigeon, which is the rock-pigeon in a very slightly altered state, has become feral in several places. Again, all recent experience shows that it is most difficult to get any wild animal to breed freely under domestication; yet on the hypothesis of the multiple origin of our pigeons, it must be assumed that at least seven or eight species were so thoroughly domesticated in ancient times by half-civilized man, as to be quite prolific under confinement.

An argument, as it seems to me, of great weight, and applicable in several other cases, is, that the above-specified breeds, though agreeing generally in constitution, habits, and in most parts of their structure, with the wild rock-pigeon, yet are certainly highly abnormal in other parts of their structure: we may look in vain throughout the whole great family of Columbidae for a beak like that of the English carrier, or that of the short-faced tumbler, or barb; for reversed feathers like those of the jacobin; for a crop like that of the pouter; for tail-feathers like those of the fantail. Hence it must be assumed not only that half-civilized man succeeded in thoroughly domesticating several species, but that he intentionally or by chance picked out extraordinarily abnormal species; and further, that these very species have since all become extinct or unknown. So many strange contingencies seem to me improbable in the highest degree.

Some facts in regard to the colouring of pigeons well deserve consideration. The rock-pigeon is of a slaty-blue, and has a white rump (the Indian sub-species, C. intermedia of Strickland, having it bluish); the tail has a terminal dark bar, with the bases of the outer feathers externally edged with white; the wings have two black bars; some semi-domestic breeds and some apparently truly wild breeds have, besides the two black bars, the wings chequered with black. These several marks do not occur together in any other species of the whole family. Now, in every one of the domestic breeds, taking thoroughly well-bred birds, all the above marks, even to the white edging of the outer tail-feathers, sometimes concur perfectly developed. Moreover, when two birds belonging to two distinct breeds are crossed, neither of which is blue or has any of the above-specified marks, the mongrel offspring are very apt suddenly to acquire these characters; for instance, I crossed some uniformly white fantails with some uniformly black barbs, and they produced mottled brown and black birds; these I again crossed together, and one grandchild of the pure white fantail and pure black barb was of as beautiful a blue colour, with the white rump, double black wing-bar, and barred and white-edged tail-feathers, as any wild rock-pigeon! We can understand these facts, on the well-known principle of reversion to ancestral characters, if all the domestic breeds have descended from the rock-pigeon. But if we deny this, we must make one of the two following highly improbable suppositions. Either, firstly, that all the several imagined aboriginal stocks were coloured and marked like the rock-pigeon, although no other existing species is thus coloured and marked, so that in each separate breed there might be a tendency to revert to the very same colours and markings. Or, secondly, that each breed, even the purest, has within a dozen or, at most, within a score of generations, been crossed by the rock-pigeon: I say within a dozen or twenty generations, for we know of no fact countenancing the belief that the child ever reverts to some one ancestor, removed by a greater number of generations. In a breed which has been crossed only once with some distinct breed, the tendency to reversion to any character derived from such cross will naturally become less and less, as in each succeeding generation there will be less of the foreign blood; but when there has been no cross with a distinct breed, and there is a tendency in both parents to revert to a character, which has been lost during some former generation, this tendency, for all that we can see to the contrary, may be transmitted undiminished for an indefinite number of generations. These two distinct cases are often confounded in treatises on inheritance.

Lastly, the hybrids or mongrels from between all the domestic breeds of pigeons are perfectly fertile. I can state this from my own observations, purposely made on the most distinct breeds. Now, it is difficult, perhaps impossible; to bring forward one case of the hybrid offspring of two animals *clearly distinct* being themselves perfectly fertile. Some authors believe that long-continued domestication eliminates this strong tendency to sterility: from the history of the dog I think there is some probability in this hypothesis, if applied to species closely related together, though it is unsupported by a single experiment. But to extend the hypothesis so far as to suppose that species, aboriginally as distinct as carriers, tumblers, pouters, and fantails now are, should yield offspring perfectly fertile, *inter se,* seems to me rash in the extreme.

From these several reasons, namely, the improbability of man having formerly got seven or eight supposed species of pigeons to breed freely under domestication; these supposed species being quite unknown in a wild state, and their becoming nowhere feral; these species having very abnormal characters in certain respects, as compared with all other Columbidae, though so like in most other respects to the rock-pigeon; the blue colour and various marks occasionally appearing in all the breeds, both when kept pure and when crossed; the mongrel offspring being perfectly fertile;—from these several reasons, taken together, I can feel no doubt that all our domestic breeds have descended from the *Columba livia* with its geographical sub-species.

In favour of this view, I may add, firstly, that *C. livia*, or the rock-pigeon, has been found capable of domestication in Europe and in India; and that it agrees in habits and in a great number of points of structure with all the domestic breeds. Secondly, although an English carrier or short-faced tumbler differs immensely in certain characters from the rock-pigeon, yet by comparing the several sub-breeds of these breeds, more especially those brought from distant countries, we can make an almost perfect series between the extremes of structure. Thirdly, those characters which are mainly distinctive of each breed, for instance the wattle and length of beak of the carrier, the shortness of that of the tumbler, and the number of tail-feathers in the fantail, are in each breed eminently variable; and the explanation of this fact will be obvious when we come to treat of selection. Fourthly, pigeons have been watched, and tended with the utmost care, and loved by many people. They have been domesticated for thousands of years in several quarters of the world; the earliest known record of pigeons is in the fifth Egyptian dynasty, about 3000 B.C, as was pointed out to me by Professor Lepsius; but Mr. Birch informs me that pigeons are given in a bill of fare in the previous dynasty. In the time of the Romans, as we hear from Pliny, immense prices were given for pigeons; "nay, they are come to this pass, that they can reckon up their pedigree and race." Pigeons were much valued by Akber Khan in India, about the year 1600; never less than 20,000 pigeons were taken with the court. "The monarchs of Iran and Turan sent him some very rare birds"; and, continues the courtly historian,

"His Majesty by crossing the breeds, which method was never practised before, has improved them astonishingly." About this same period the Dutch were as eager about pigeons as were the old Romans. The paramount importance of these considerations in explaining the immense amount of variation which pigeons have undergone, will be obvious when we treat of Selection. We shall then, also, see how it is that the breeds so often have a somewhat monstrous character. It is also a most favourable circumstance for the production of distinct breeds, that male and female pigeons can be easily mated for life; and thus different breeds can be kept together in the same aviary.

I have discussed the probable origin of domestic pigeons at some, yet quite insufficient, length; because when I first kept pigeons and watched the several kinds, knowing well how true they bred, I felt fully as much difficulty in believing that they could ever have descended from a common parent, as any naturalist could in coming to a similar conclusion in regard to the many species of finches, or other large groups of birds, in nature. One circumstance has struck me much; namely, that all the breeders of the various domestic animals and the cultivators of plants, with whom I have ever conversed, or whose treatises I have read, are firmly convinced that the several breeds to which each has attended, are descended from so many aboriginally distinct species. Ask, as I have asked, a celebrated raiser of Hereford cattle, whether his cattle might not have descended from longhorns, and he will laugh you to scorn. I have never met a pigeon, or poultry, or duck, or rabbit fancier, who was not fully convinced that each main breed was descended from a distinct species. Van Mons, in his treatise on pears and apples, shows how utterly he disbelieves that the several sorts, for instance a Ribston-pippin or Codlin-apple, could ever have proceeded from the seeds of the same tree. Innumerable other examples could be given. The explanation, I think, is simple: from long-continued study they are strongly impressed with the differences between the several races; and though they well know that each race varies slightly, for they win their prizes by selecting such slight differences, yet they ignore all general arguments, and refuse to sum up in their minds slight differences accumulated during many successive generations. May not those naturalists who, knowing far less of the laws of inheritance than does the breeder, and knowing no more than he does of the intermediate links in the long lines of descent, yet admit that many of our domestic races have descended from the same parents—may they not learn a lesson of caution, when they deride the idea of species in a state of nature being lineal descendants of other species?

Selection. Let us now briefly consider the steps by which domestic races have been produced, either from one or from several allied species. Some little effect may, perhaps, be attributed to the direct action of the external conditions of life, and some little to habit; but he would be a bold man who would account by such agencies for the differences of a dray and race horse, a greyhound and bloodhound, a carrier and tumbler pigeon. One of the most remarkable features in our domesticated races is that we see in them adaptation, not indeed to the animal's or plant's own good, but to man's use or fancy. Some variations useful to him have probably arisen suddenly, or by one step; many botanists, for instance, believe that the fuller's teazle, with its hooks, which cannot be rivalled by any mechanical contrivance, is only a variety of the wild Dipsacus; and this amount of change may have suddenly arisen in a seedling. So it has probably been with the turnspit dog; and this is known to have been the case with the ancon sheep. But when we compare the dray-horse and race-horse, the dromedary and camel, the various breeds of sheep fitted either for cultivated land or mountain pasture, with the wool of one breed good for one purpose, and that of another breed for another purpose; when we compare the many breeds of dogs, each good for man in very different ways; when we compare the game-cock, so pertinacious in battle, with other breeds so little quarrelsome, with "everlasting layers" which never desire to sit, and with the bantam so small and elegant; when we compare the host of agricultural, culinary, orchard, and flower-garden races of plants, most useful to man at different seasons and for different purposes, or so beautiful in his eyes, we must, I think, look further than to mere variability. We cannot suppose that all the breeds were suddenly produced as perfect and as useful as we now see them; indeed, in several cases, we know that this has not been their history. The key is man's power of accumulative selection: nature gives successive variations; man adds them up in certain directions useful to him. In this sense he may be said to make for himself useful breeds.

The great power of this principle of selection is not hypothetical. It is certain that several of our eminent breeders have, even within a single lifetime, modified to a large extent some breeds of cattle and sheep. In order fully to realize what they have done, it is almost necessary to read several of the many treatises devoted to this subject, and to inspect the animals. Breeders habitually speak of an animal's organisation as something quite plastic, which they can model almost as they please. If I had space I could quote numerous passages to this effect from highly competent authorities. Youatt, who was probably better acquainted with the works of agriculturalists than almost any other individual, and who was himself a very good judge of an animal, speaks of the principle of selection as "that which enables the agriculturist, not only to modify the character of his flock, but to change it altogether. It is the magician's wand, by means of which he may summon into life whatever form and mould he pleases." Lord Somerville, speaking of what breeders have done for sheep, says:—"It would seem as if they had chalked out upon a wall a form perfect in itself, and then had given it existence." That most skilful breeder, Sir John Sebright, used to say, with respect to pigeons, that "he would produce any given feather in three years, but it would take him six years to obtain head and beak." In Saxony the importance of the principle of selection in regard to merino sheep is so fully recognised, that men follow it as a trade: the sheep are placed on a table and are studied, like a picture by a connoisseur; this is done three times at intervals of months, and the sheep are each time marked and classed, so that the very best may ultimately be selected for breeding.

What English breeders have actually effected is proved by the enormous prices given for animals with a good pedigree; and these have now been exported to almost every quarter of the world. The improvement is by no means generally due to crossing different breeds; all the best breeders are strongly opposed to this practice, except sometimes amongst closely allied sub-breeds. And when a cross has been made, the closest selection is far more indispensable even than in ordinary cases. If selection consisted merely in separating some very distinct variety, and breeding from it, the principle would be so obvious as hardly to be worth notice; but its importance consists in the great effect produced by the accumulation in one direction, during successive generations, of differences absolutely inappreciable by an uneducated eye—differences which I for one have vainly attempted to appreciate. Not one man in a thousand has accuracy of eye and judgement sufficient to become an eminent breeder. If gifted with these qualities, and he studies his subject for years, and devotes his lifetime to it with indomitable perseverance, he will succeed, and may make great improvements; if he wants any of these qualities, he will assuredly fail. Few would readily believe in the natural capacity and years of practice requisite to become even a skilful pigeon-fancier.

* * *

In the case of animals with separate sexes, facility in preventing crosses is an important element of success in the formation of new races,—at least, in a country which is already stocked with other races. In this respect enclosure of the land plays a part. Wandering savages or the inhabitants of open plains rarely possess more than one breed of the same species. Pigeons can be mated for life, and this is a great convenience to the fancier, for thus many races may be kept true, though mingled in the same aviary; and this circumstance must have largely favoured the improvement and formation of new breeds. Pigeons, I may add, can be propagated in great numbers and at a very quick rate, and inferior birds may be freely rejected, as when killed they serve for food. On the other hand, cats, from their nocturnal rambling habits, cannot be matched, and, although so much valued by women and children, we hardly ever see a distinct breed kept up; such breeds as we do sometimes see are almost always imported from some other country, often from islands. Although I do not doubt that some domestic animals vary less than others, yet the rarity or absence of distinct breeds of the cat, the donkey, peacock, goose, &c., may be attributed in main part to selection not having been brought into play: in cats, from the difficulty in pairing them; in donkeys, from only a few being kept by poor people, and little attention paid to their breeding; in peacocks, from not being very easily reared and a large stock not kept; in geese, from being valuable only for two purposes, food and feathers, and more especially from no pleasure having been felt in the display of distinct breeds.

To sum up on the origin of our Domestic Races of animals and plants. I believe that the conditions of life, from their action on the reproductive system, are so far of the highest importance as causing variability. I do not believe that variability is an inherent and necessary contingency, under all circumstances, with all organic beings, as some authors have thought. The effects of variability are modified by various degrees of inheritance and of reversion. Variability is governed by many unknown laws, more especially by that of correlation of growth. Something may be attributed to the direct action of the conditions of life. Something must be attributed to use and disuse. The final result is thus rendered infinitely complex. In some cases, I do not doubt that the intercrossing of species, aboriginally distinct, has played an important part in the origin of our domestic productions. When in any country several domestic breeds have once been established, their occasional intercrossing, with the aid of selection, has, no doubt, largely aided in the formation of new sub-breeds; but the importance of the crossing of varieties has, I believe, been greatly exaggerated, both in regard to animals and to those plants which are propagated by seed. In plants which are temporarily propagated by cuttings, buds, &c., the importance of the crossing both of distinct species and of varieties is immense; for the cultivator here quite disregards the extreme variability both of hybrids and mongrels, and the frequent sterility of hybrids; but the cases of plants not propagated by seed are of little importance to us, for their endurance is only temporary. Over all these causes of Change I am convinced that the accumulative action of Selection, whether applied methodically and more quickly, or unconsciously and more slowly, but more efficiently, is by far the predominant Power.

Editor's Introduction to Chapter III

In Chapter I Darwin claimed that domesticated animals and plants vary more than do wild ones, a claim that advanced his arguments about human modification of domesticated forms, but which left him with a problem concerning animals and plants in the wild: Is there enough variation in nature to support his doctrine that continual modification of a geographical race of one species can result in a new species—a small four-toed grazing animal, for example, being transformed into a horse?

Darwin faced this question in his Chapter II, entitled *Variation Under Nature*, arguing that there is indeed sufficient variation in nature to make such transformations possible. His argument there took a peculiar and interesting indirect course. I have not included Chapter II in the present selections, but here is a brief summary of its findings.

Variation in Nature

Darwin pointed out that experts were often in disagreement whether a particular population was simply a geographical race of some species, or a separate species in itself. Experts were universally agreed that there *were* separate species of animals and plants in the world, but they could not reach agreement about identifying these species—nor even about what it meant to *be* a species. From this lack of agreement Darwin concluded that nature exhibits a tremendous degree of variation—enough to generate all the races, varieties, and sub-varieties that trouble the lives of biologists. It is to this argument that Darwin alludes in the opening paragraph of the present Chapter III, *Struggle for Existence.*

Struggle for Existence

The struggle for existence is the central concept in Darwin's argument, notwithstanding that Darwin himself asserted that *natural selection* is the main principle (see, for example, pages 7 and 26, above). For in order for natural selection to operate, the living world must have a certain competitive character. Darwin presents that character in Chapter III, and it is here that he parts company with almost all thinkers who had already written about the origin of species.

Although Chapter III presents no technical difficulties, its rhetoric is sufficiently striking that a bit of guidance may be useful. Darwin's central image in this chapter is the *face of nature*. This image demands close attention. Initially, Darwin will characterize nature's face as "bright with gladness" (page 32)—the struggle for existence, with its attendant death and destruction, being likewise presented as going on beneath this face and concealed by it. But as Darwin's picture of nature continues to develop, the struggle for existence becomes more and more prominent. Finally the face of nature appears, in one of Darwin's most astonishing figures, as a surface continually being penetrated by thousands of sharp wedges (page 35), incessantly being driven into it by force. With this image, then, even the surface appearance of serenity in nature has been removed.

Darwin admits that he uses the phrase *struggle for existence* with a "large and metaphorical" meaning that encompasses many cases in which living things cannot be said to be "struggling" in any ordinary sense. Fundamentally, Darwin explains, it is the *mutual dependence of living things on one another* that he has in mind. Why, then, does he insist on calling such interdependence a "struggle" for existence? The reason is the tendency of all organic beings to reproduce at so high a rate that only a small fraction of them can survive—and therefore all forms of interdependence actually become forms of competition. This fact, when put together with the tendency of living things to vary, results in a process of selection that is natural rather than a reflection of conscious human purpose.

The interdependence of living things, Darwin believes, is so intense that almost no differences among them are neutral. Practically all variations affect both survival and ability to leave offspring. However the consequences of these modifications usually require so much time to manifest themselves that the face of nature rarely reveals what is going on underneath. Instead, it gives us a false image of permanence and serenity.

For Darwin, there are two additional and perhaps surprising consequences of this new view of the living world. In the penultimate paragraph of the present chapter he asks us to imagine what would happen if we conveyed a whole population of animals or plants to a new environment. Even if climatic conditions were not much changed, the transported population would be subject to new and different conditions of life because they would have to deal with the native animals and plants already present in the new environment. They would probably need to change, therefore, in order to live well. In the chapter's final paragraph Darwin asks us to imagine what those changes might be; and he makes the sobering claim that in no case could we anticipate what changes would be beneficial, because our knowledge of the mutual relations of organic beings is so meager. This is very important because it raises the question, how complete a knowledge of nature can we have under the conditions of a struggle for existence? More baldly stated, is a *science* of living nature possible?

The chapter's final paragraph also raises the question how we as human beings, whether scientists or not, should feel about this new vision of nature. Darwin knows that we will find it disturbing, and he is obviously troubled himself. Think carefully about his words of consolation at the end of the chapter. This is one of many places in the book where it is clear that Darwin means to address not just scientists, but all thoughtful readers.

CHAPTER III

STRUGGLE FOR EXISTENCE

BEFORE entering on the subject of this chapter, I must make a few preliminary remarks, to show how the struggle for existence bears on Natural Selection. It has been seen in the last chapter[11] that amongst organic beings in a state of nature there is some individual variability; indeed I am not aware that this has ever been disputed. It is immaterial for us whether a multitude of doubtful forms be called species or sub-species or varieties; what rank, for instance, the two or three hundred doubtful forms of British plants are entitled to hold, if the existence of any well-marked varieties be admitted. But the mere existence of individual variability and of some few well-marked varieties, though necessary as the foundation for the work, helps us but little in understanding how species arise in nature. How have all those exquisite adaptations of one part of the organisation to another part, and to the conditions of life, and of one distinct organic being to another being, been perfected? We see these beautiful co-adaptations most plainly in the woodpecker and misseltoe; and only a little less plainly in the humblest parasite which clings to the hairs of a quadruped or feathers of a bird; in the structure of the beetle which dives through the water; in the plumed seed which is wafted by the gentlest breeze; in short, we see beautiful adaptations everywhere and in every part of the organic world.

Again, it may be asked, how is it that varieties, which I have called incipient species, become ultimately converted into good and distinct species, which in most cases obviously differ from each other far more than do the varieties of the same species? How do those groups of species, which constitute what are called distinct genera, and which differ from each other more than do the species of the same genus, arise? All these results, as we shall more fully see in the next chapter, follow inevitably from the struggle for life. Owing to this struggle for life, any variation, however slight and from whatever cause proceeding, if it be in any degree profitable to an individual of any species, in its infinitely complex relations to other organic beings and to external nature, will tend to the preservation of that individual, and will generally be inherited by its offspring. The offspring, also, will thus have a better chance of surviving, for, of the many individuals of any species which are periodically born, but a small number can survive. I have called this principle, by which each slight variation, if useful, is preserved, by the term of Natural Selection, in order to mark its relation to man's power of selection. We have seen that man by selection

11. *the last chapter*: Darwin refers to Chapter II, which is not included in our selections but is summarized in the Editor's Introduction to the present chapter.

But we have better evidence on this subject than mere theoretical calculations, namely, the numerous recorded cases of the astonishingly rapid increase of various animals in a state of nature, when circumstances have been favourable to them during two or three following seasons. Still more striking is the evidence from our domestic animals of many kinds which have run wild in several parts of the world: if the statements of the rate of increase of slow-breeding cattle and horses in South America, and latterly in Australia, had not been well authenticated, they would have been quite incredible. So it is with plants: cases could be given of introduced plants which have become common throughout whole islands in a period of less than ten years. Several of the plants now most numerous over the wide plains of La Plata, clothing square leagues of surface almost to the exclusion of all other plants, have been introduced from Europe; and there are plants which now range in India, as I hear from Dr. Falconer, from Cape Comorin to the Himalaya, which have been imported from America since its discovery. In such cases, and endless instances could be given, no one supposes that the fertility of these animals or plants has been suddenly and temporarily increased in any sensible degree. The obvious explanation is that the conditions of life have been very favourable, and that there has consequently been less destruction of the old and young, and that nearly all the young have been enabled to breed. In such cases the geometrical ratio of increase, the result of which never fails to be surprising, simply explains the extraordinarily rapid increase and wide diffusion of naturalised productions in their new homes.

In a state of nature almost every plant produces seed, and amongst animals there are very few which do not annually pair. Hence we may confidently assert, that all plants and animals are tending to increase at a geometrical ratio, that all would most rapidly stock every station in which they could any how exist, and that the geometrical tendency to increase must be checked by destruction at some period of life. Our familiarity with the larger domestic animals tends, I think, to mislead us: we see no great destruction falling on them, and we forget that thousands are annually slaughtered for food, and that in a state of nature an equal number would have somehow to be disposed of.

The only difference between organisms which annually produce eggs or seeds by the thousand, and those which produce extremely few, is, that the slow-breeders would require a few more years to people, under favourable conditions, a whole district, let it be ever so large. The condor lays a couple of eggs and the ostrich a score, and yet in the same country the condor may be the more numerous of the two: the Fulmar petrel lays but one egg, yet it is believed to be the most numerous bird in the world. One fly deposits hundreds of eggs, and another, like the hippobosca, a single one; but this difference does not determine how many individuals of the two species can be supported in a district. A large number of eggs is of some importance to those species, which depend on a rapidly fluctuating amount of food, for it allows them rapidly to increase in number. But the real importance of a large number of eggs or seeds is to make up for much destruction at some period of life; and this period in the great majority of cases is an early one. If an animal can in any way protect its own eggs or young, a small number may be produced, and yet the average stock be fully kept up; but if many eggs or young are destroyed, many must be produced, or the species will become extinct. It would suffice to keep up the full number of a tree, which lived on an average for a thousand years, if a single seed were produced once in a thousand years, supposing that this seed were never destroyed, and could be ensured to germinate in a fitting place. So that in all cases, the average number of any animal or plant depends only indirectly on the number of its eggs or seeds.

In looking at Nature, it is most necessary to keep the foregoing considerations always in mind—never to forget that every single organic being around us may be said to be striving to the utmost to increase in numbers; that each lives by a struggle at some period of its life; that heavy destruction inevitably falls either on the young or old, during each generation or at recurrent intervals. Lighten any check, mitigate the destruction ever so little, and the number of the species will almost instantaneously increase to any amount. The face of Nature may be compared to a yielding surface, with ten thousand sharp wedges packed close together and driven inwards by incessant blows, sometimes one wedge being struck, and then another with grealer force.

What checks the natural tendency of each species to increase in number is most obscure. Look at the most vigorous species; by as much as it swarms in numbers, by so much will its tendency to increase be still further increased. We know not exactly what the checks are in even one single instance. Nor will this surprise any one who reflects how ignorant we are on this head, even in regard to mankind, so incomparably better known than any other animal. This subject has been ably treated by several authors, and I shall, in my future work, discuss some of the checks at considerable length, more especially in regard to the feral animals of South America. Here I will make only a few remarks, just to recall to the reader's mind some of the chief points. Eggs or very young animals seem generally to suffer most, but this is not invariably the case. With plants there is a vast destruction of seeds, but, from some observations which I have made, I believe that it is the seedlings which suffer most from germinating in ground already thickly stocked with other plants. Seedlings, also, are destroyed in vast numbers by various enemies; for instance, on a piece of ground three feet long and two wide, dug and cleared, and where there could be no choking from other plants, I marked all the seedlings of our native weeds as they came up; and out of the 357 no less than 295 were destroyed, chiefly by slugs and insects. If turf which has long been mown, and the case would be the same with turf closely browsed by quadrupeds, be let to grow, the more vigorous plants gradually kill the less vigorous, though fully grown, plants: thus out of twenty species growing on a little plot of turf (three feet by four) nine species perished from the other species being allowed to grow up freely.

* * *

Climate plays an important part in determining the average numbers of a species, and periodical seasons of extreme cold or drought, I believe to be the most effective of all checks. I estimated that the winter of 1854–55 destroyed four-fifths of the birds in my own grounds; and this is a tremendous destruction, when we remember that ten per cent. is an extraordinarily severe mortality from epidemics with man. The action of climate seems at first sight to be quite independent of the struggle for existence; but in so far as climate chiefly acts in reducing food, it brings on the most severe struggle between the individuals, whether of the same or of distinct species, which subsist on the same kind of food. Even when climate, for instance extreme cold, acts directly, it will be the least vigorous, or those which have got least food through the advancing winter, which will suffer most. When we travel from south to north, or from a damp region to a dry, we invariably see some species gradually getting rarer and rarer, and finally disappearing; and the change of climate being conspicuous, we are tempted to attribute the whole effect to its direct action. But this is a very false view: we forget that each species, even where it most abounds, is constantly suffering enormous destruction at some period of its life, from enemies or from competitors for the same place and food; and if these enemies or competitors be in the least degree favoured by any slight change of climate, they will increase in numbers, and, as each area is already fully stocked with inhabitants, the other species will decrease. When we travel southward and see a species decreasing in numbers, we may feel sure that the cause lies quite as much in other species being favoured, as in this one being hurt. So it is when we travel northward, but in a somewhat lesser degree, for the number of species of all kinds, and therefore of competitors, decreases northwards; hence in going northward, or in ascending a mountain, we far oftener meet with stunted forms, due to the *directly* injurious action of climate, than we do in proceeding southwards or in descending a mountain. When we reach the Arctic regions, or snow-capped summits, or absolute deserts, the struggle for life is almost exclusively with the elements.

That climate acts in main part indirectly by favouring other species, we may clearly see in the prodigious number of plants in our gardens which can perfectly well endure our climate, but which never become naturalised, for they cannot compete with our native plants, nor resist destruction by our native animals.

* * *

Many cases are on record showing how complex and unexpected are the checks and relations between organic beings, which have to struggle together in the same country. I will give only a single instance, which, though a simple one, has interested me. In Staffordshire, on the estate of a relation where I had ample means of investigation, there was a large and extremely barren heath, which had never been touched by the hand of man; but several hundred acres of exactly the same nature had been enclosed twenty-five years previously and planted with Scotch fir. The change in the native vegetation of the planted part of the heath was most remarkable, more than is generally seen in passing from one quite different soil to another: not only the proportional numbers of the heath-plants were wholly changed, but twelve species of plants (not counting grasses and carices) flourished in the plantations, which could not be found on the heath. The effect on the insects must have been still greater, for six insectivorous birds were very common in the plantations, which were not to be seen on the heath; and the heath was frequented by two or three distinct insectivorous birds. Here we see how potent has been the effect of the introduction of a single tree, nothing whatever else having been done, with the exception that the land had been enclosed, so that cattle could not enter. But how important an element enclosure is, I plainly saw near Farnham, in Surrey. Here there are extensive heaths, with a few clumps of old Scotch firs on the distant hill-tops: within the last ten years large spaces have been enclosed, and self-sown firs are now springing up in multitudes, so close together that all cannot live. When I ascertained that these young trees had not been sown or planted, I was so much surprised at their numbers that I went to several points of view, whence I could examine hundreds of acres of the unenclosed heath, and literally I could not see a single Scotch fir, except the old planted clumps. But on looking closely between the stems of the heath, I found a multitude of seedlings and little trees, which had been perpetually browsed down by the cattle. In one square yard, at a point some hundreds yards distant from one of the old clumps, I counted thirty-two little trees;

and one of them, judging from the rings of growth, had during twenty-six years tried to raise its head above the stems of the heath, and had failed. No wonder that, as soon as the land was enclosed, it became thickly clothed with vigorously growing young firs. Yet the heath was so extremely barren and so extensive that no one would ever have imagined that cattle would have so closely and effectually searched it for food.

Here we see that cattle absolutely determine the existence of the Scotch fir; but in several parts of the world insects determine the existence of cattle. Perhaps Paraguay offers the most curious instance of this; for here neither cattle nor horses nor dogs have ever run wild, though they swarm southward and northward in a feral state; and Azara and Rengger have shown that this is caused by the greater number in Paraguay of a certain fly, which lays its eggs in the navels of these animals when first born. The increase of these flies, numerous as they are, must be habitually checked by some means, probably by birds. Hence, if certain insectivorous birds (whose numbers are probably regulated by hawks or beasts of prey) were to increase in Paraguay, the flies would decrease—then cattle and horses would become feral, and this would certainly greatly alter (as indeed I have observed in parts of South America) the vegetation: this again would largely affect the insects; and this, as we just have seen in Staffordshire, the insectivorous birds, and so onwards in ever-increasing circles of complexity. We began this series by insectivorous birds, and we have ended with them. Not that in nature the relations can ever be as simple as this. Battle within battle must ever be recurring with varying success; and yet in the long-run the forces are so nicely balanced, that the face of nature remains uniform for long periods of time, though assuredly the merest trifle would often give the victory to one organic being over another. Nevertheless so profound is our ignorance, and so high our presumption, that we marvel when we hear of the extinction of an organic being; and as we do not see the cause, we invoke cataclysms to desolate the world, or invent laws on the duration of the forms of life!

I am tempted to give one more instance showing how plants and animals, most remote in the scale of nature, are bound together by a web of complex relations. I shall hereafter have occasion to show that the exotic *Lobelia fulgens*, in this part of England, is never visited by insects, and consequently, from its peculiar structure, never can set a seed. Many of our orchidaceous plants absolutely require the visits of moths to remove their pollen-masses and thus to fertilise them. I have, also, reason to believe that humble-bees are indispensable to the fertilisation of the heartsease (*Viola tricolor*), for other bees do not visit this flower. From experiments which I have tried, I have found that the visits of bees, if not indispensable, are at least highly beneficial to the fertilisation of our clovers; but humble-bees alone visit the common red clover (*Trifolium pratense*), as other bees cannot reach the nectar. Hence I have very little doubt, that if the whole genus of humble-bees became extinct or very rare in England, the heartsease and red clover would become very rare, or wholly disappear. The number of humble-bees in any district depends in a great degree on the number of field-mice, which destroy their combs and nests; and Mr. H. Newman, who has long attended to the habits of humble-bees, believes that "more than two thirds of them are thus destroyed all over England." Now the number of mice is largely dependent, as every one knows, on the number of cats; and Mr. Newman says, "Near villages and small towns I have found the nests of humble-bees more numerous than elsewhere, which I attribute to the number of cats that destroy the mice." Hence it is quite credible that the presence of a feline animal in large numbers in a district might determine, through the intervention first of mice and then of bees, the frequency of certain flowers in that district!

In the case of every species, many different checks, acting at different periods of life, and during different seasons or years, probably come into play; some one check or some few being generally the most potent, but all concurring in determining the average number or even the existence of the species. In some cases it can be shown that widely-different checks act on the same species in different districts. When we look at the plants and bushes clothing an entangled bank, we are tempted to attribute their proportional numbers and kinds to what we call chance. But how false a view is this! Every one has heard that when an American forest is cut down, a very different vegetation springs up; but it has been observed that the trees now growing on the ancient Indian mounds, in the Southern United States, display the same beautiful diversity and proportion of kinds as in the surrounding virgin forests. What a struggle between the several kinds of trees must here have gone on during long centuries, each annually scattering its seeds by the thousand; what war between insect and insect—between insects, snails, and other animals with birds and beasts of prey—all striving to increase, and all feeding on each other or on the trees or their seeds and seedlings, or on the other plants which first clothed the ground and thus checked the growth of the trees! Throw up a handful of feathers, and all must fall to the ground according to definite laws; but how simple is this problem compared to the action and reaction of the innumerable plants and animals which have determined, in the course of centuries, the proportional numbers and kinds of trees now growing on the old Indian ruins!

The dependency of one organic being on another, as of a parasite on its prey, lies generally between beings remote in the scale of nature. This is often the case with those which may strictly be said to struggle with each other for existence, as in the case of locusts and grass-feeding quadrupeds. But the struggle almost invariably will be most severe between the individuals of the same species, for they frequent the same districts, require the same food, and are exposed to the same dangers. In the case of varieties of the same species, the struggle will generally be almost equally severe, and we sometimes see the contest soon decided: for instance, if several varieties of wheat be sown together, and the mixed seed be resown, some of the varieties which best suit the soil or climate, or are naturally the most fertile, will beat the others and so yield more seed, and will consequently in a few years quite supplant the other varieties. To keep up a mixed stock of even such extremely close varieties as the variously coloured sweet-peas, they must be each year harvested separately, and the seed then mixed in due proportion, otherwise the weaker kinds will steadily decrease in numbers and disappear. So again with the varieties of sheep: it has been asserted that certain mountain-varieties will starve out other mountain-varieties, so that they cannot be kept together. The same result has followed from keeping together different varieties of the medicinal leech. It may even be doubted whether the varieties of any one of our domestic plants or animals have so exactly the same strength, habits, and constitution, that the original proportions of a mixed stock could be kept up for half a dozen generations, if they were allowed to struggle together, like beings in a state of nature, and if the seed or young were not annually sorted.

* * *

A corollary of the highest importance may be deduced from the foregoing remarks, namely, that the structure of every organic being is related, in the most essential yet often hidden manner, to that of all other organic beings, with which it comes into competition for food or residence, or from which it has to escape, or on which it preys. This is obvious in the structure of the teeth and talons of the tiger; and in that of the legs and claws of the parasite which clings to the hair on the tiger's body. But in the beautifully plumed seed of the dandelion, and in the flattened and fringed legs of the water-beetle, the relation seems at first confined to the elements of air and water. Yet the advantage of plumed seeds no doubt stands in the closest relation to the land being already thickly clothed by other plants; so that the seeds may be widely distributed and fall on unoccupied ground. In the water-beetle, the structure of its legs, so well adapted for diving, allows it to compete with other aquatic insects, to hunt for its own prey, and to escape serving as prey to other animals.

The store of nutriment laid up within the seeds of many plants seems at first sight to have no sort of relation to other plants. But from the strong growth of young plants produced from such seeds (as peas and beans), when sown in the midst of long grass, I suspect that the chief use of the nutriment in the seed is to favour the growth of the young seedling, whilst struggling with other plants growing vigorously all around.

Look at a plant in the midst of its range, why does it not double or quadruple its numbers? We know that it can perfectly well withstand a little more heat or cold, dampness or dryness, for elsewhere it ranges into slightly hotter or colder, damper or drier districts. In this case we can clearly see that if we wished in imagination to give the plant the power of increasing in number, we should have to give it some advantage over its competitors, or over the animals which preyed on it. On the confines of its geographical range, a change of constitution with respect to climate would clearly be an advantage to our plant; but we have reason to believe that only a few plants or animals range so far, that they are destroyed by the rigour of the climate alone. Not until we reach the extreme confines of life, in the arctic regions or on the borders of an utter desert, will competition cease. The land may be extremely cold or dry, yet there will be competition between some few species, or between the individuals of the same species, for the warmest or dampest spots.

Hence, also, we can see that when a plant or animal is placed in a new country amongst new competitors, though the climate may be exactly the same as in its former home, yet the conditions of its life will generally be changed in an essential manner. If we wished to increase its average numbers in its new home, we should have to modify it in a different way to what we should have done in its native country; for we should have to give it some advantage over a different set of competitors or enemies.

It is good thus to try in our imagination to give any form some advantage over another. Probably in no single instance should we know what to do, so as to succeed. It will convince us of our ignorance on the mutual relations of all organic beings; a conviction as necessary, as it seems to be difficult to acquire. All that we can do, is to keep steadily in mind that each organic being is striving to increase at a geometrical ratio; that each at some period of its life, during some season of the year, during each generation or at intervals, has to struggle for life, and to suffer great destruction. When we reflect on this struggle, we may console ourselves with the full belief, that the war of nature is not incessant, that no fear is felt, that death is generally prompt, and that the vigorous, the healthy, and the happy survive and multiply.

Editor's Introduction to Chapter IV

In Chapter III Darwin put forward the idea that the living world is characterized by the struggle for existence. Now in Chapter IV he will show us how to think about such a world. It has been said that Darwin's chief strength as a scientist was his allegiance to the facts rather than to unsubstantiated speculation. Although there is much truth in that statement—indeed Darwin himself repeatedly stresses the value of careful observation—it should not blind us to the highly theoretical, and even speculative, character of this chapter. We should not expect a series of examples of natural selection designed to win us over to his theory on purely empirical grounds. Even if Darwin had wanted to proceed in that way, he could not have done so, for such examples do not exist—or, at least, were not known to Darwin. This fact was already foreshadowed near the end of Chapter III, where he challenged us to state how we would modify a species in order to give it some advantage over another species, and concluded that in no case could we possibly know what to change. Later on, in our introduction to Chapter XI, we will have more to say concerning instances of natural selection in the world of our experience. For the present we should notice that all the examples of natural selection in this chapter are, as Darwin repeatedly acknowledges, *imaginary* ones. It is astonishing how many readers overlook this fact, despite Darwin's explicit statements. Darwin will choose these imagined examples with great care, for they are to be his main vehicle for teaching us how to think about the world, insofar as it is governed by the struggle for existence.

Illustrations of the theory of natural selection

Darwin will begin with two such imagined cases, the first one concerning wolves, the second involving flowers and insects. These examples stand at the extremes of a series running from the simple to the complex. Two factors make Darwin's wolf example a very simple one. First, the element of competition is easy to see, for the wolves are directly competing with one another for deer. Second, the environment—including the deer that are the wolf's main prey—does not change. Under these circumstances the question is entirely and exclusively about the *wolves*: will those wolves that become more adept at catching deer prevail and pass on their traits to their offspring, eventually changing the character of the wolf population? If the changes are great enough we may no longer be willing to call them wolves, giving them instead another name—they will have become a new species.

Darwin then considers a further possibility—that a wolf population might divide into two species, one hunting in the highlands, the other in the lowlands. So we now have a picture, admittedly a very simplified one, of how divergence of character might come about. Darwin does offer some examples of divergent populations of wolves that have been reported in the Catskills. What is not said here is as important as what is. Even if there do exist two kinds of wolves in the Catskills, that fact by itself does not show how the two kinds came about. The mere fact of their existence does not tell us whether they are the descendants of favored geographical races of an original wolf species, or if they arose in some other way.

At the complex end of the series stands Darwin's second example, which concerns bees and the flowers they feed from. In order to appreciate it you will need to know something abo the structure of a flower in relati to an insect. Examine the flow depicted here; it is hermaphroditi meaning that it has both male a female parts. The *stamen,* consi ing of anther and filament, is t male structure; pollen is produc on the anther. The stigma, sty and ovary together are called the *pistil*. This is the female structure; eggs are produced in the ovary.

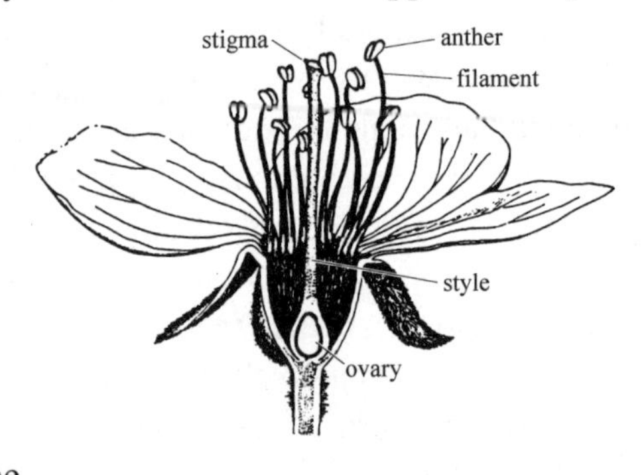

The surface of the stigma is sticky. Pollen grains first stick to the stigma, then are transported down the style into the ovary where they fertilize the eggs. *Nectar*, which is greatly coveted by insects, is produced by glandlike structures either around the ovary or at the base of the petals or the base of the stamens. In order to reach the nector, an insect must first pass the stamens. Depending both on its own shape and on that of the flower, it may brush against the anthers and stigma. When it contacts the anthers it picks up pollen on its hairy body. When it visits the next flower some of this pollen will be transferred to the stigma, effecting cross-fertilization.

Darwin's second example is highly complex. For one thing, there is no stable background against which changes may be understood; rather, both bee and flower are changing with respect to one another. Moreover, the struggle for life is more difficult to discern. Here insect does not compete with flower, nor does flower compete with the insect, in any ordinary sense—on the contrary, both are in some sense cooperating to make their structures fit together for mutual benefit. Instead, different insects can be said to be competing with one another to make themselves conformable to the flowers, while different flowers may be said to be competing with one another to make themselves attractive to the insects.

Deeper issues are also at stake. The flowers on these plants, with the help of the insects, gradually become incapable of self-fertilization, even though the flowers continue to be hermaphroditic in structure. This is good for the species because intercrossing is generally beneficial; but it also makes the life of the plant more precarious because visits by insects, instead of being merely useful for reproduction, ultimately become *essential*. More and more, bee and flower become intimately connected with and mutually dependent upon one another.

Intercrossing

For Darwin, the word *intercrossing* covers a variety of phenomena. It can refer to individuals in closely related populations breeding with one another on the borders of their ranges, as in Darwin's example of two kinds of wolves in the Catskills. It can also apply, more dramatically, to individuals of different species breeding with one another, as when horse and donkey cross to produce mule. Finally, it can even refer to hermaphroditic flowers on the same plant eschewing self-fertilization, crossing instead with other flowers.

The matter is intricate, but two limiting factors stand out. First, a certain amount of intercrossing appears to be necessary. If, for example, a population is too small, breeding will occur frequently between close relatives, leading eventually to loss of vigor in the stock. The reasons for this are complex and were not fully understood by Darwin; but a need for intercrossing would explain why plants with hermaphroditic flowers tend either to develop single-sex flowers, or else to lose the power of self-fertilization and come to depend on insects or the wind for cross-fertilization. On the other hand, too much intercrossing can blur the direction of a population's development. If a population is varying in a certain way, there must be some restriction on how extensively that population can breed outside itself; otherwise it will not be able to retain the direction in which it is varying. In order for a population to evolve extensively, therefore, it will have to be both fairly large and somewhat isolated.

Isolation

Is geographical isolation necessary for a species to diverge into two or more species? This is a fundamental and problematic issue for Darwin. Although Darwin did not see his way to a clear solution, he did recognize all its essential elements. Imagine a species that is distributed over a large area, but one so subdivided by geographical barriers (such as mountains and rivers) that the species is, in effect, subdivided into a number of discrete populations. Then it is easy to see how each of these subgroups could evolve differently from the others, especially if the conditions of life in the several areas are so different that the organisms in each area are forced into different ways of life. Each of these areas is then a kind of island; and in these circumstances it is easy to see how divergence of character can occur.

The more difficult case, and the one Darwin cannot quite think through, is when a species is distributed over a wide range with conditions of life that gradually change throughout the area, but where there are no geographical barriers to define different regions. In this case, two competing things are happening. First, organisms in different portions of the range have a tendency to adapt to their respective conditions of life, and thus to diverge from one another. On the other hand, since there are no geographical barriers between regions, there will always be some interbreeding between groups. Such interbreeding mixes the groups together, constantly undermining their incipient divergence into different species.

Darwin wondered whether the differing ways of life that emerge between different groups can themselves become barriers to interbreeding. For example, might the two populations of wolves in the Catskills, as a consequence of hunting in different ways, develop different breeding habits, and thus become uncongenial to one another? And in general, are there cases where incipient species change can bring about reproductive isolation, and thereby promote more intense change? This issue is very important to Darwin, for he believes that the most vigorous kinds of animals and plants are those which are created under the conditions of intense competition found in large continental areas, and that forms created under conditions of complete geographical isolation lacked vigor. To give one important example, native Australian plants and animals, which had developed in complete isolation from life on the great continents, were regularly supplanted by Eurasian species. Darwin never solved the problem to his satisfaction, and it remains a controversial topic among evolutionary biologists today. He did, however, state the question clearly and deeply.

Divergence of character

In the penultimate section of this chapter Darwin will explore the phenomenon he calls *divergence of character*, which he represents graphically in the diagram on pages 61 and 62. There, each cluster of branching lines represents the production of several incipient new species from an existing one. As the diagram shows, most of these new lines of descent become extinct very soon; but a few persist in time and themselves become the origin of more diversity. Darwin's diagram may be thought of as a highly abstract picture of the Tree of Life, mentioned earlier in the Editor's General Introduction. The branching of the Tree of Life represents the living world as continually coming into existence; it is not a static image but a highly dynamic one.

CHAPTER IV

NATURAL SELECTION

HOW will the struggle for existence, discussed too briefly in the last chapter, act in regard to variation? Can the principle of selection, which we have seen is so potent in the hands of man, apply in nature? I think we shall see that it can act most effectually. Let it be borne in mind in what an endless number of strange peculiarities our domestic productions, and, in a lesser degree, those under nature, vary; and how strong the hereditary tendency is. Under domestication, it may be truly said that the whole organisation becomes in some degree plastic. Let it be borne in mind how infinitely complex and close-fitting are the mutual relations of all organic beings to each other and to their physical conditions of life. Can it, then, be thought improbable, seeing that variations useful to man have undoubtedly occurred, that other variations useful in some way to each being in the great and complex battle of life, should sometimes occur in the course of thousands of generations? If such do occur, can we doubt (remembering that many more individuals are born than can possibly survive) that individuals having any advantage, however slight, over others, would have the best chance of surviving and of procreating their kind? On the other hand, we may feel sure that any variation in the least degree injurious would be rigidly destroyed. This preservation of favourable variations and the rejection of injurious variations, I call Natural Selection. Variations neither useful nor injurious would not be affected by natural selection, and would be left a fluctuating element, as perhaps we see in the species called polymorphic.

We shall best understand the probable course of natural selection by taking the case of a country undergoing some physical change, for instance, of climate. The proportional numbers of its inhabitants would almost immediately undergo a change, and some species might become extinct. We may conclude, from what we have seen of the intimate and complex manner in which the inhabitants of each country are bound together, that any change in the numerical proportions of some of the inhabitants, independently of the change of climate itself, would most seriously affect many of the others. If the country were open on its borders, new forms would certainly immigrate, and this also would seriously disturb the relations of some of the former inhabitants. Let it be remembered how powerful the influence of a single introduced tree or mammal has been shown to be. But in the case of an island, or of a country partly surrounded by barriers, into which new and better adapted forms could not freely enter, we should then have places in the economy of nature which would assuredly be better filled up, if some of the original inhabitants were in some manner modified; for, had the area been open to immigration, these same places would have been seized on by intruders. In such case, every slight modification, which in the course of ages chanced to arise, and which in any way favoured the individuals of any of the species, by better adapting them to their altered conditions, would tend to be preserved; and natural selection would thus have free scope for the work of improvement.

We have reason to believe, as stated in the first chapter, that a change in the conditions of life, by specially acting on the reproductive system, causes or increases variability; and in the foregoing case the conditions of life are supposed to have undergone a change, and this would manifestly be favourable to natural selection, by giving a better chance of profitable variations occurring; and unless profitable variations do occur, natural selection can do nothing. Not that, as I believe, any extreme amount of variability is necessary; as man can certainly produce great results by adding up in any given direction mere individual differences, so could Nature, but far more easily, from having incomparably longer time at her disposal. Nor do I believe that any great physical change, as of climate, or any unusual degree of isolation to check immigration, is actually necessary to produce new and unoccupied places for natural selection to fill up by modifying and improving some of the varying inhabitants. For as all the inhabitants of each country are struggling together with nicely balanced forces, extremely slight modifications in the structure or habits of one inhabitant would often give it an advantage over others; and still further modifications of the same kind would often still further increase the advantage. No country can be named in which all the native inhabitants are now so perfectly adapted to each other and to the physical conditions under which they live, that none of them could anyhow be improved; for in all countries, the natives have been so far conquered by naturalised productions, that they have allowed foreigners to take firm possession of the land. And as foreigners have thus everywhere beaten some of the natives, we may safely conclude that the natives might have been modified with advantage, so as to have better resisted such intruders.

As man can produce and certainly has produced a great result by his methodical and unconscious means of selection, what may not nature effect? Man can act only on external and visible characters: nature cares nothing for appearances, except in so far as they may be useful to any being. She can act on every internal organ, on every shade of constitutional difference, on the whole machinery of life. Man selects only for his own good; Nature only for that of the being which she tends. Every selected character is fully exercised by her; and the being is placed under well-suited conditions of life. Man keeps the natives of many climates in the same country; he seldom exercises each selected character in some peculiar and fitting manner; he feeds a long and a short beaked pigeon on the same food; he does not exercise a long-backed or long-legged quadruped in any peculiar manner; he exposes sheep with long and short wool to the same climate. He does not allow the most vigorous males to struggle for the females. He does not rigidly destroy all inferior animals, but protects during each varying season, as far as lies in his power, all his productions. He often begins his selection by some half-monstrous form; or at least by some modification prominent enough to catch his eye, or to be plainly useful to him. Under nature, the slightest difference of structure or constitution may well turn the nicely-balanced scale in the struggle for life, and so be preserved. How fleeting are the wishes and efforts of man! how short his time! and consequently how poor will his products be, compared with those accumulated by nature during whole geological periods. Can we wonder, then, that nature's productions should be far "truer" in character than man's productions; that they should be infinitely better adapted to the most complex conditions of life, and should plainly bear the stamp of far higher workmanship?

It may be said that natural selection is daily and hourly scrutinising, throughout the world, every variation, even the slightest; rejecting that which is bad, preserving and adding up all that is good; silently and insensibly working, whenever and wherever opportunity offers, at the improvement of each organic being in relation to its organic and inorganic conditions of life. We see nothing of these slow changes in progress, until the hand of time has marked the long lapses of ages, and then so imperfect is our view into long past geological ages, that we only see that the forms of life are now different from what they formerly were.

* * *

Natural selection will modify the structure of the young in relation to the parent, and of the parent in relation to the young. In social animals it will adapt the structure of each individual for the benefit of the community; if each in consequence profits by the selected change. What natural selection cannot do, is to modify the structure of one species, without giving it any advantage, for the good of another species; and though statements to this effect may be found in works of natural history, I cannot find one case which will bear investigation. A structure used only once in an animal's whole life, if of high importance to it, might be modified to any extent by natural selection; for instance, the great jaws possessed by certain insects, and used exclusively for opening the cocoon—or the hard tip to the beak of nestling birds, used for breaking the egg. It has been asserted, that of the best short-beaked tumbler-pigeons more perish in the egg than are able to get out of it; so that fanciers assist in the act of hatching. Now, if nature had to make the beak of a full-grown pigeon very short for the bird's own advantage, the process of modification would be very slow, and there would be simultaneously the most rigorous selection of the young birds within the egg, which had the most powerful and hardest beaks, for all with weak beaks would inevitably perish: or, more delicate and more easily broken shells might be selected, the thickness of the shell being known to vary like every other structure.

* * *

Illustrations of the action of Natural Selection. In order to make it clear how, as I believe, natural selection acts, I must beg permission to give one or two imaginary illustrations. Let us take the case of a wolf, which preys on various animals, securing some by craft, some by strength, and some by fleetness; and let us suppose that the fleetest prey, a deer for instance, had from any change in the country increased in numbers, or that other prey had decreased in numbers, during that season of the year when the wolf is hardest pressed for food. I can under such circumstances see no reason to doubt that the swiftest and slimmest wolves would have the best chance of surviving, and so be preserved or selected,—provided always that they retained strength to master their prey at this or at some other period of the year, when they might be compelled to prey on other animals. I can see no more reason to doubt this, than that man can improve the fleetness of his greyhounds by careful and methodical selection, or by that unconscious selection which results from each man trying to keep the best dogs without any thought of modifying the breed.

Even without any change in the proportional numbers of the animals on which our wolf preyed, a cub might be born with an innate tendency to pursue certain kinds of prey. Nor can this be thought very improbable; for we often observe great differences in the natural tendencies of our domestic animals; one cat, for instance, taking to catch rats, another mice; one cat, according to Mr. St. John, bringing home winged game, another hares or rabbits, and another hunting on marshy ground and almost nightly catching woodcocks or snipes. The tendency to catch rats rather than mice is known to be inherited. Now, if any slight innate change of habit or of structure benefited an individual wolf, it would have the best chance of surviving and of leaving offspring. Some of its young would probably inherit the same habits or structure, and by the repetition of this process, a new variety might be formed which would either supplant or coexist with the parent-form of wolf. Or, again, the wolves inhabiting a mountainous district, and those frequenting the lowlands, would naturally be forced to hunt different prey; and from the continued preservation of the individuals best fitted for the two sites, two varieties might slowly be formed. These varieties would cross and blend where they met; but to this subject of intercrossing we shall soon have to return. I may add, that, according to Mr. Pierce, there are two varieties of the wolf inhabiting the Catskill Mountains in the United States, one with a light greyhound-like form, which pursues deer, and the other more bulky, with shorter legs, which more frequently attacks the shepherd's flocks.

Let us now take a more complex case. Certain plants excrete a sweet juice, apparently for the sake of eliminating something injurious from their sap: this is effected by glands at the base of the stipules in some Leguminosae, and at the back of the leaf of the common laurel. This juice, though small in quantity, is greedily sought by insects. Let us now suppose a little sweet juice or nectar to be excreted by the inner bases of the petals of a flower. In this case insects in seeking the nectar would get dusted with pollen, and would certainly often transport the pollen from one flower to the stigma of another flower. The flowers of two distinct individuals of the same species would thus get crossed; and the act of crossing, we have good reason to believe (as will hereafter be more fully alluded to), would produce very vigorous seedlings, which consequently would have the best chance of flourishing and surviving. Some of these seedlings would probably inherit the nectar-excreting power. Those individual flowers which had the largest glands or nectaries, and which excreted most nectar, would be oftenest visited by insects, and would be oftenest crossed; and so in the long-run would gain the upper hand. Those flowers, also, which had their stamens and pistils placed, in relation to the size and habits of the particular insects which visited them, so as to favour in any degree the transportal of their pollen from flower to flower, would likewise be favoured or selected. We might have taken the case of insects visiting flowers for the sake of collecting pollen instead of nectar; and as pollen is formed for the sole object of fertilisation, its destruction appears a simple loss to the plant; yet if a little pollen were carried, at first occasionally and then habitually, by the pollen-devouring insects from flower to flower, and a cross thus effected, although nine-tenths of the pollen were destroyed, it might still be a great gain to the plant; and those individuals which produced more and more pollen, and had larger and larger anthers, would be selected.

When our plant, by this process of the continued preservation or natural selection of more and more attractive flowers, had been rendered highly attractive to insects, they would, unintentionally on their part, regularly carry pollen from flower to flower; and that they can most effectually do this, I could easily show by many striking instances. I will give only one—not as a very striking case, but as likewise illustrating one step in the separation of the sexes of plants, presently to be alluded to. Some holly-trees bear only male flowers, which have four stamens producing rather a small quantity of pollen, and a rudimentary pistil; other holly-trees bear only female flowers; these have a full-sized pistil, and four stamens with shrivelled anthers, in which not a grain of pollen can be detected. Having found a female tree exactly sixty yards from a male tree, I put the stigmas of twenty flowers, taken from different branches, under the microscope, and on all, without exception, there were pollen-grains, and on some a profusion of pollen. As the wind had set for several days from the female to the male tree, the pollen could not thus have been carried. The weather had been cold and boisterous, and therefore not favourable to bees, nevertheless every female flower which I examined had been effectually fertilised by the bees, accidentally dusted with pollen, having flown from tree to tree in search of nectar. But to return to our imaginary case: as soon as the plant had been rendered so highly attractive to insects that pollen was regularly carried from flower to flower, another process might commence. No naturalist doubts the advantage of what has been called the "physiological division of labour," hence we may believe that it would be advantageous to a plant to produce stamens alone in one flower or on one whole plant, and pistils alone in another flower or on another plant. In plants under culture and placed under new conditions of life, sometimes the male organs and sometimes the female organs become more or less impotent; now if we suppose this to occur in ever so slight a degree under nature, then as pollen is already carried regularly from flower to flower, and as a more complete separation of the sexes of our plant would be advantageous on the principle of the division of labour, individuals with this tendency more and more increased, would be continually favoured or selected, until at last a complete separation of the sexes would be effected.

Let us now turn to the nectar-feeding insects in our imaginary case: we may suppose the plant of which we have been slowly increasing the nectar by continued selection, to be a common plant; and that certain insects depended in main part on its nectar for food. I could give many facts, showing how anxious bees are to save time; for instance, their habit of cutting holes and sucking the nectar at the bases of certain flowers, which they can, with a very little more trouble, enter by the mouth. Bearing such facts in mind, I can see no reason to doubt that an accidental deviation in the size and form of the body, or in the curvature and length of the proboscis, &c., far too slight to be appreciated by us, might profit a bee or other insect, so that an individual so characterised would be able to obtain its food more quickly, and so have a better chance of living and leaving descendants. Its descendants would probably inherit a tendency to a similar slight deviation of structure. The tubes of the corollas of the common red and incarnate clovers (*Trifolium pratense* and *incarnatum*) do not on a hasty glance appear to differ in length; yet the hive-bee can easily suck the nectar out of the incarnate clover, but not out of the common red clover, which is visited by humble-bees alone; so that whole fields of the red clover offer in vain an abundant supply of precious nectar to the hive-bee. Thus it might be a great advantage to the hive-bee to have a slightly longer or differently constructed proboscis. On the other hand, I have found by experiment that the fertility of clover greatly depends on bees visiting and moving parts of the corolla, so as to push the pollen on to the stigmatic surface. Hence, again, if humble-bees were to become rare in any country, it might be a great advantage to the red clover to have a shorter or more deeply divided tube to its corolla, so that the hive-bee could visit its flowers. Thus I can understand how a flower and a bee might slowly become, either simultaneously or one after the other, modified and adapted in the most perfect manner to each other, by the continued preservation of individuals presenting mutual and slightly favourable deviations of structure.

I am well aware that this doctrine of natural selection, exemplified in the above imaginary instances, is open to the same objections which were at first urged against Sir Charles Lyell's noble views[15] on "the modern changes of the earth, as illustrative of geology"; but we now very seldom hear the action, for instance, of the coast-waves, called a trifling and insignificant cause, when applied to the excavation of gigantic valleys or to the formation of the longest lines of inland cliffs. Natural selection can act only by the preservation and accumulation of infinitesimally small inherited modifications, each profitable to the preserved being; and as modern geology has almost banished such views as the excavation of a great valley by a single diluvial wave, so will natural selection, if it be a true principle, banish the belief of the continued creation of new organic beings, or of any

15. Lyell's "noble views" were those presented in his *Principles of Geology* (page 32*n*, above). Among them was his understanding of geologic changes as having come about through the accumulation of minute alterations over enormous periods of time—a view that was attacked by writers who believed that huge changes could be explained only by large, catastrophic causes.

great and sudden modification in their structure.

* * *

Circumstances favourable to Natural Selection. This is an extremely intricate subject. A large amount of inheritable and diversified variability is favourable, but I believe mere individual differences suffice for the work. A large number of individuals, by giving a better chance for the appearance within any given period of profitable variations, will compensate for a lesser amount of variability in each individual, and is, I believe, an extremely important element of success. Though nature grants vast periods of time for the work of natural selection, she does not grant an indefinite period; for as all organic beings are striving, it may be said, to seize on each place in the economy of nature, if any one species does not become modified and improved in a corresponding degree with its competitors, it will soon be exterminated.

In man's methodical selection, a breeder selects for some definite object, and free intercrossing will wholly stop his work. But when many men, without intending to alter the breed, have a nearly common standard of perfection, and all try to get and breed from the best animals, much improvement and modification surely but slowly follow from this unconscious process of selection, notwithstanding a large amount of crossing with inferior animals. Thus it will be in nature; for within a confined area, with some place in its polity not so perfectly occupied as might be, natural selection will always tend to preserve all the individuals varying in the right direction, though in different degrees, so as better to fill up the unoccupied place. But if the area be large, its several districts will almost certainly present different conditions of life; and then if natural selection be modifying and improving a species in the several districts, there will be intercrossing with the other individuals of the same species on the confines of each. And in this case the effects of intercrossing can hardly be counterbalanced by natural selection always tending to modify all the individuals in each district in exactly the same manner to the conditions of each; for in a continuous area, the conditions will generally graduate away insensibly from one district to another. The intercrossing will most affect those animals which unite for each birth, which wander much, and which do not breed at a very quick rate. Hence in animals of this nature, for instance in birds, varieties will generally be confined to separated countries; and this I believe to be the case. In hermaphrodite organisms which cross only occasionally, and likewise in animals which unite for each birth, but which wander little and which can increase at a very rapid rate, a new and improved variety might be quickly formed on any one spot, and might there maintain itself in a body, so that whatever intercrossing took place would be chiefly between the individuals of the same new variety. A local variety when once thus formed might subsequently slowly spread to other districts. On the above principle, nurserymen always prefer getting seed from a large body of plants of the same variety, as the chance of intercrossing with other varieties is thus lessened.

Even in the case of slow-breeding animals, which unite for each birth, we must not overrate the effects of intercrosses in retarding natural selection; for I can bring a considerable catalogue of facts, showing that within the same area, varieties of the same animal can long remain distinct, from haunting different stations, from breeding at slightly different seasons, or from varieties of the same kind preferring to pair together.

Intercrossing plays a very important part in nature in keeping the individuals of the same species, or of the same variety, true and uniform in character. It will obviously thus act far more efficiently with those animals which unite for each birth; but I have already attempted to show that we have reason to believe that occasional intercrosses take place with all animals and with all plants. Even if these take place only at long intervals, I am convinced that the young thus produced will gain so much in vigour and fertility over the offspring from long-continued self-fertilisation, that they will have a better chance of surviving and propagating their kind; and thus, in the long run, the influence of intercrosses, even at rare intervals, will be great. If there exist organic beings which never intercross, uniformity of character can be retained amongst them, as long as their conditions of life remain the same, only through the principle of inheritance, and through natural selection destroying any which depart from the proper type; but if their conditions of life change and they undergo modification, uniformity of character can be given to their modified offspring, solely by natural selection preserving the same favourable variations.

Isolation, also, is an important element in the process of natural selection. In a confined or isolated area, if not very large, the organic and inorganic conditions of life will generally be in a great degree uniform; so that natural selection will tend to modify all the individuals of a varying species throughout the area in the same manner in relation to the same conditions. Intercrosses, also, with the individuals of the same species, which otherwise would have inhabited the surrounding and differently circumstanced districts, will be prevented. But isolation probably acts more efficiently in checking the immigration of better adapted organisms, after any physical change, such as of climate or elevation of the land, &c.; and thus new places in the natural economy of the country are left open for the old inhabitants to struggle for, and become adapted to, through modifications in their structure and constitution. Lastly, isolation, by checking immigration and consequently competition, will give time for any new variety to be slowly improved; and this may sometimes be of importance in the production of new species. If, however, an isolated area be very small, either from being surrounded by barriers, or from having very peculiar physical conditions, the total number of the individuals supported on it will necessarily be very small; and fewness of individuals will greatly retard the production of new species through natural selection, by decreasing the chance of the appearance of favourable variations.

If we turn to nature to test the truth of these remarks, and look at any small isolated area, such as an oceanic island, although the total number of the species inhabiting it, will be found to be small, as we shall see in our chapter on geographical distribution; yet of these species a very large proportion are endemic,—that is, have been produced there, and nowhere else. Hence an oceanic island at first sight seems to have been highly favourable for the production of new species. But we may thus greatly deceive ourselves, for to ascertain whether a small isolated area, or a large open area like a continent, has been most favourable for the production of new organic forms, we ought to make the comparison within equal times; and this we are incapable of doing.

Although I do not doubt that isolation is of considerable importance in the production of new species, on the whole I am inclined to believe that largeness of area is of more importance, more especially in the production of species, which will prove capable of enduring for a long period, and of spreading widely. Throughout a great and open area, not only will there be a better chance of favourable variations arising from the large number of individuals of the same species there supported, but the conditions of life are infinitely complex from the large number of already existing species; and if some of these many species become modified and improved, others will have to be improved in a corresponding degree or they will be exterminated. Each new form, also, as soon as it has been much improved, will be able to spread over the open and continuous area, and will thus come into competition with many others. Hence more new places will be formed, and the competition to fill them will be more severe, on a large than on a small and isolated area. Moreover, great areas, though now continuous, owing to oscillations of level, will often have recently existed in a broken condition, so that the good effects of isolation will generally, to a certain extent, have concurred. Finally, I conclude that, although small isolated areas probably have been in some respects highly favourable for the production of new species, yet that the course of modification will generally have been more rapid on large areas; and what is more important, that the new forms produced on large areas, which already have been victorious over many competitors, will be those that will spread most widely, will give rise to most new varieties and species, and will thus play an important part in the changing history of the organic world.

We can, perhaps, on these views, understand some facts which will be again alluded to in our chapter on geographical distribution; for instance, that the productions of the smaller continent of Australia have formerly yielded, and apparently are now yielding, before those of the larger Europaeo-Asiatic area. Thus, also, it is that continental productions have everywhere become so largely naturalised on islands. On a small island, the race for life will have been less severe, and there will have been less modification and less extermination. Hence, perhaps, it comes that the flora of Madeira, according to Oswald Heer, resembles the extinct tertiary flora of Europe. All fresh-water basins, taken together, make a small area compared with that of the sea or of the land; and, consequently, the competition between fresh-water productions will have been less severe than elsewhere; new forms will have been more slowly formed, and old forms more slowly exterminated. And it is in fresh water that we find seven genera of Ganoid fishes, remnants of a once preponderant order: and in fresh water we find some of the most anomalous forms now known in the world, as the Ornithorhynchus and Lepidosiren, which, like fossils, connect to a certain extent orders now widely separated in the natural scale. These anomalous forms may almost be called living fossils; they have endured to the present day, from having inhabited a confined area, and from having thus been exposed to less severe competition.

To sum up the circumstances favourable and unfavourable to natural selection, as far as the extreme intricacy of the subject permits. I conclude, looking to the future, that for terrestrial productions a large continental area, which will probably undergo many oscillations of level, and which consequently will exist for long periods in a broken condition, will be the most favourable for the production of many new forms of life, likely to endure long and to spread widely. For the area will first have existed as a continent, and the inhabitants, at this period numerous in individuals and kinds, will have been subjected to very severe competition. When converted by subsidence into large separate islands, there will still exist many individuals of the same species on each island: intercrossing on the confines of the range of each species will thus be checked: after physical changes of any kind, immigration will be prevented, so that new places in the polity of each island will have to be filled up by modifications of the old inhabitants; and time will be allowed for the varieties in each to become well modified and perfected. When, by renewed elevation, the islands shall be re-converted into a continental area, there will again be severe competition: the most favoured or improved varieties will be enabled to spread: there will be much extinction of the less improved forms, and the relative proportional numbers of the various inhabitants of the renewed continent will again be changed; and again there will be a fair field for natural selection to improve still further the inhabitants, and thus produce new species.

That natural selection will always act with extreme slowness, I fully admit. Its action depends on there being places in the polity of nature, which can be better occupied by some of the inhabitants of the country undergoing modification of some kind. The existence of such places will often depend on physical changes, which are generally very slow, and on the immigration of better adapted forms having been checked. But the action of natural selection will probably still oftener depend on some of the inhabitants becoming slowly modified; the mutual relations of many of the other inhabitants being thus disturbed. Nothing can be effected, unless favourable variations occur, and variation itself is apparently always a very slow process. The process will often be greatly retarded by free intercrossing. Many will exclaim that these several causes are amply sufficient wholly to stop the action of natural selection. I do not believe so. On the other hand, I do believe that natural selection will always act very slowly, often only at long intervals of time, and generally on only a very few of the inhabitants of the same region at the same time. I further believe, that this very slow, intermittent action of natural selection accords perfectly well with what geology tells us of the rate and manner at which the inhabitants of this world have changed.

Slow though the process of selection may be, if feeble man can do much by his powers of artificial selection, I can see no limit to the amount of change, to the beauty and infinite complexity of the coadaptations between all organic beings, one with another and with their physical conditions of life, which may be effected in the long course of time by nature's power of selection.

* * *

Divergence of Character. The principle, which I have designated by this term, is of high importance on my theory, and explains, as I believe, several important facts. In the first place, varieties, even strongly-marked ones, though having somewhat of the character of species—as is shown by the hopeless doubts in many cases how to rank them—yet certainly differ from each other far less than do good and distinct species. Nevertheless, according to my view, varieties are species in the process of formation, or are, as I have called them, incipient species. How, then, does the lesser difference between varieties become augmented into the greater difference between species? That this does habitually happen, we must infer from most of the innumerable species throughout nature presenting well-marked differences; whereas varieties, the supposed prototypes and parents of future well-marked species, present slight and ill-defined differences. Mere chance, as we may call it, might cause one variety to differ in some character from its parents, and the offspring of this variety again to differ from its parent in the very tame character and in a greater degree; but this alone would never account for so habitual and large an amount of differencc as that between varietics of the same species and species of the same genus.

As has always been my practice, let us seek light on this head from our domestic productions. We shall here find something analogous. A fancier is struck by a pigeon having a slightly shorter beak; another fancier is struck by a pigeon having a rather longer beak; and on the acknowledged principle that "fanciers do not and will not admire a medium standard, but like extremes," they both go on (as has actually occurred with tumbler-pigeons) choosing and breeding from birds with longer and longer beaks, or with shorter and shorter beaks. Again, we may suppose that at an early period one man preferred swifter horses; another stronger and more bulky horses. The early differences would be very slight; in the course of time, from the continued selection of swifter horses by some breeders, and of stronger ones by others, the differences would become greater, and would be noted as forming two sub-breeds; finally, after the lapse of centuries, the sub-breeds would become converted into two well-established and distinct breeds. As the differences slowly become greater, the inferior animals with intermediate characters, being neither very swift nor very strong, will have been neglected, and will have tended to disappear. Here, then, we see in man's productions the action of what may be called the principle of divergence, causing differences, at first barely appreciable, steadily to increase, and the breeds to diverge in character both from each other and from their common parent.

But how, it may be asked, can any analogous principle apply in nature? I believe it can and does apply most efficiently, from the simple circumstance that the more diversified the descendants from any one species become in structure, constitution, and habits, by so much will they be better enabled to seize on many and widely diversified places in the polity of nature, and so be enabled to increase in numbers.

We can clearly see this in the case of animals with simple habits. Take the case of a carnivorous quadruped, of which the number that can be supported in any country has long ago arrived at its full average. If its natural powers of increase be allowed to act, it can succeed in increasing (the country not undergoing any change in its conditions) only by its varying descendants seizing on places at present occupied by other animals: some of them, for instance, being enabled to feed on new kinds of prey, either dead or alive; some inhabiting new stations, climbing trees, frequenting water, and some perhaps becoming less carnivorous. The more diversified in habits and structure the descendants of our carnivorous animal became, the more places they would be enabled to occupy. What applies to one animal will apply throughout all time to all animals—that is, if they vary—for otherwise natural selection can do nothing. So it will be with plants. It has been experimentally proved, that if a plot of ground be sown with one species of grass, and a similar plot be sown with several distinct genera of grasses, a greater number of plants and a greater weight of dry herbage can thus be raised. The same has been found to hold good when first one variety and then several mixed varieties of wheat have been sown on equal spaces of ground. Hence, if any one species of grass were to go on varying, and those varieties were continually selected which differed from each other in at all the same manner as distinct species and genera of grasses differ from each other, a greater number of individual plants of this species of grass, including its modified descendants, would succeed in living on the same piece of ground. And we well know that each species and each variety of grass is annually sowing almost countless seeds; and thus, as it may be said, is striving its utmost to increase its numbers. Consequently, I cannot doubt that in the course of many thousands of generations, the most distinct varieties of any one species of grass would always have the best chance of succeeding and of increasing in numbers, and thus of supplanting the less distinct varieties; and varieties, when rendered very distinct from each other, take the rank of species.

The truth of the principle, that the greatest amount of life can be supported by great diversification of structure, is seen under many natural circumstances. In an extremely small area, especially if freely open to immigration, and where the contest between individual and individual must be severe, we always find great diversity in its inhabitants. For instance, I found that a piece of turf, three feet by four in size, which had been exposed for many years to exactly the same conditions, supported twenty species of plants, and these belonged to eighteen genera and to eight orders, which shows how much these plants differed from each other. So it is with the plants and insects on small and uniform islets; and so in small ponds of fresh water. Farmers find that they can raise most food by a rotation of plants belonging to the most different orders: nature follows what may be called a simultaneous rotation. Most of the animals and plants which live close round any small piece of ground, could live on it (supposing it not to be in any way peculiar in its nature), and may be said to be striving to the

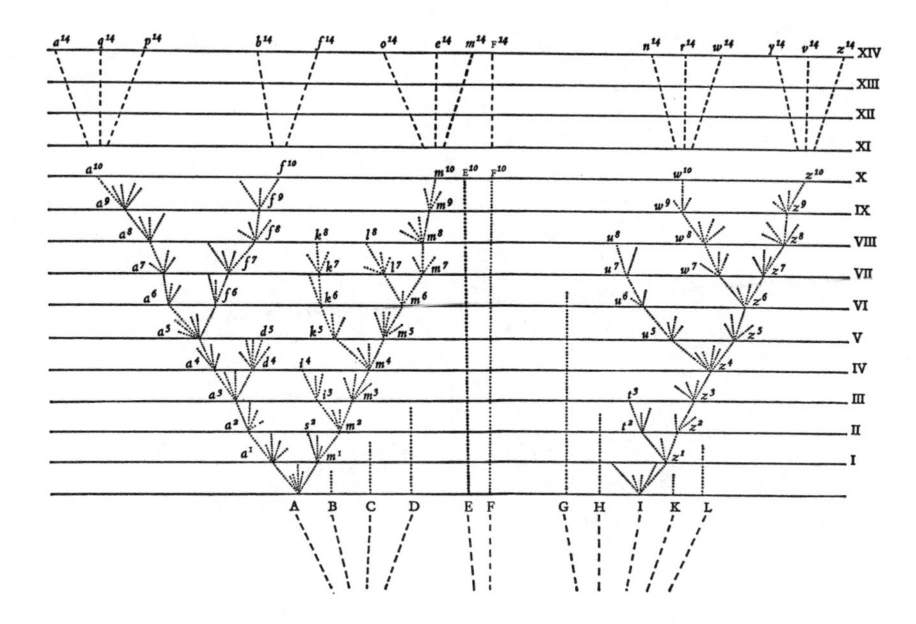

utmost to live there; but, it is seen, that where they come into the closest competition with each other, the advantages of diversification of structure, with the accompanying differences of habit and constitution, determine that the inhabitants, which thus jostle each other most closely, shall, as a general rule, belong to what we call different genera and orders.

* * *

The accompanying diagram will aid us in understanding this rather perplexing subject. Let A to L represent the species of a genus large in its own country; these species are supposed to resemble each other in unequal degrees, as is so generally the case in nature, and as is represented in the diagram by the letters standing at unequal distances. I have said a large genus, because we have seen in the second chapter, that on an average more of the species of large genera vary than of small genera; and the varying species of the large genera present a greater number of varieties. We have, also, seen that the species, which are the commonest and the most widely-diffused, vary more than rare species with restricted ranges. Let (A) be a common, widely-diffused, and varying species, belonging to a genus large in its own country. The little fan of diverging dotted lines of unequal lengths proceeding from (A), may represent its varying offspring. The variations are supposed to be extremely slight, but of the most diversified nature; they are not supposed all to appear simultaneously, but often after long intervals of time; nor are they all supposed to endure for equal periods. Only those variations which are in some way profitable will be preserved

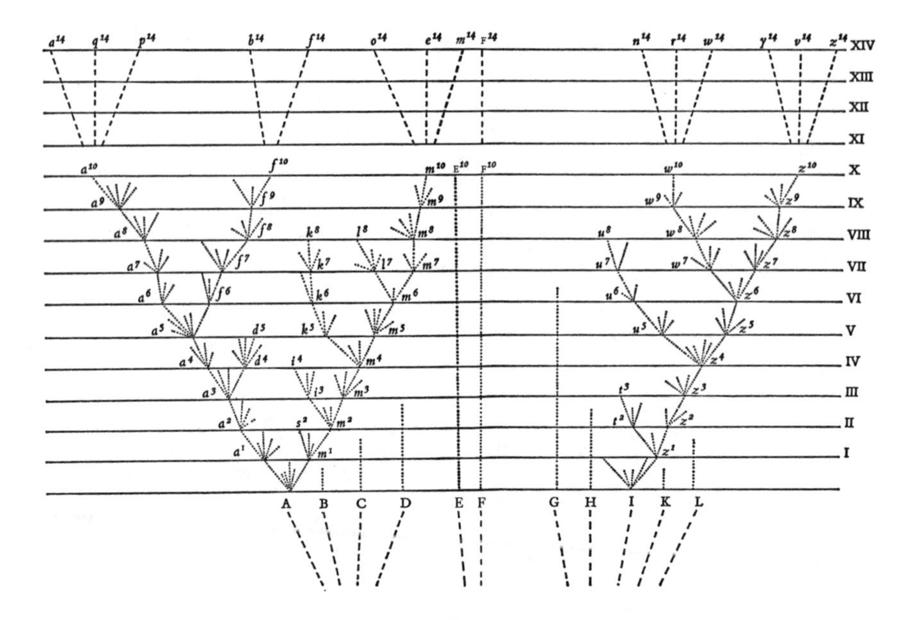

or naturally selected. And here the importance of the principle of benefit being derived from divergence of character comes in; for this will generally lead to the most different or divergent variations (represented by the outer dotted lines) being preserved and accumulated by natural selection. When a dotted line reaches one of the horizontal lines, and is there marked by a small numbered letter, a sufficient amount of variation is supposed to have been accumulated to have formed a fairly well-marked variety, such as would be thought worthy of record in a systematic work.

The intervals between the horizontal lines in the diagram, may represent each a thousand generations; but it would have been better if each had represented ten thousand generations. After a thousand generations, species (A) is supposed to have produced two fairly well-marked varieties, namely a^1 and m^1. These two varieties will generally continue to be exposed to the same conditions which made their parents variable, and the tendency to variability is in itself hereditary, consequently they will tend to vary, and generally to vary in nearly the same manner as their parents varied. Moreover, these two varieties, being only slightly modified forms, will tend to inherit those advantages which made their common parent (A) more numerous than most of the other inhabitants of the same country; they will likewise partake of those more general advantages which made the genus to which the parent-species belonged, a large genus in its own country. And these circumstances we know to be favourable to the production of new varieties.

If, then, these two varieties be variable, the most divergent of their variations will generally be preserved during the next thousand generations. And after this interval, variety a^1 is supposed in the diagram to have produced variety a^2, which will, owing to the principle of divergence, differ more from (A) than did variety a^1. Variety m^1 is supposed to have produced two varieties, namely m^2 and s^2, differing from each other, and more considerably from their common parent (A). We may continue the process by similar steps for any length of time; some of the varieties, after each thousand generations, producing only a single variety, but in a more and more modified condition, some producing two or three varieties, and some failing to produce any. Thus the varieties or modified descendants, proceeding from the common parent (A), will generally go on increasing in number and diverging in character. In the diagram the process is represented up to the ten-thousandth generation, and under a condensed and simplified form up to the fourteen-thousandth generation.

But I must here remark that I do not suppose that the process ever goes on so regularly as is represented in the diagram, though in itself made somewhat irregular. I am far from thinking that the most divergent varieties will invariably prevail and multiply: a medium form may often long endure, and may or may not produce more than one modified descendant; for natural selection will always act according to the nature of the places which are either unoccupied or not perfectly occupied by other beings; and this will depend on infinitely complex relations. But as a general rule, the more diversified in structure the descendants from any one species can be rendered, the more places they will be enabled to seize on, and the more their modified progeny will be increased. In our diagram the line of succession is broken at regular intervals by small numbered letters marking the successive forms which have become sufficiently distinct to be recorded as varieties. But these breaks are imaginary, and might have been inserted anywhere, after intervals long enough to have allowed the accumulation of a considerable amount of divergent variation.

* * *

Summary of Chapter. If during the long course of ages and under varying conditions of life, organic beings vary at all in the several parts of their organisation, and I think this cannot be disputed; if there be, owing to the high geometrical powers of increase of each species, at some age, season, or year, a severe struggle for life, and this certainly cannot be disputed; then, considering the infinite complexity of the relations of all organic beings to each other and to their conditions of existence, causing an infinite diversity in structure, constitution, and habits, to be advantageous to them, I think it would be a most extraordinary fact if no variation ever had occurred useful to each being's own welfare, in the same way as so many variations have occurred useful to man. But if variations useful to any organic being do occur, assuredly individuals thus characterised will have the best chance of being preserved in the struggle for life; and from the strong principle of inheritance they will tend to produce offspring similarly characterised. This principle of preservation, I have called, for the sake of brevity, Natural Selection. Natural selection, on the principle of qualities being inherited at corresponding ages, can modify the egg, seed, or young, as easily as the adult. Amongst many animals, sexual selection will give its aid to ordinary selection, by assuring to the most vigorous and best adapted males the greatest number of offspring. Sexual selection will also give characters useful to the males alone, in their struggles with other males.

Whether natural selection has really thus acted in nature, in modifying and adapting the various forms of life to their several conditions and stations, must be judged of by the general tenour and balance of evidence given in the following chapters. But we already see how it entails extinction; and how largely extinction has acted in the world's history, geology plainly declares. Natural selection, also, leads to divergence of character; for more living beings can be supported on the same area the more they diverge in structure, habits, and constitution, of which we see proof by looking at the inhabitants of any small spot or at naturalised productions. Therefore during the modification of the descendants of any one species, and during the incessant struggle of all species to increase in numbers, the more diversified these descendants become, the better will be their chance of succeeding in the battle of life. Thus the small differences distinguishing varieties of the same species, will steadily tend to increase till they come to equal the greater differences between species of the same genus, or even of distinct genera.

We have seen that it is the common, the widely-diffused, and widely-ranging species, belonging to the larger genera, which vary most; and these will tend to transmit to their modified offspring that superiority which now makes them dominant in their own countries. Natural selection, as has just been remarked, leads to divergence of character and to much extinction of the less improved and intermediate forms of life. On these principles, I believe, the nature of the affinities of all organic beings may be explained. It is a truly wonderful fact—the wonder of which we are apt to overlook from familiarity—that all animals and all plants throughout all time and space should he related to each other in group subordinate to group, in the manner which we everywhere behold—namely, varieties of the same species most closely related together, species of the same genus less closely and unequally related together, forming sections and sub-genera, species of distinct genera much less closely related, and genera related in different degrees, forming subfamilies, families, orders, sub-classes, and classes. The several subordinate groups in any class cannot be ranked in a single file, but seem rather to be clustered round points, and these round other points, and so on in almost endless cycles. On the view that each species has been independently created, I can see no explanation of this great fact in the classification of all organic beings; but, to the best of my judgment, it is explained through inheritance and the complex action of natural selection, entailing extinction and divergence of character, as we have seen illustrated in the diagram.

The affinities of all the beings of the same class have sometimes been represented by a great tree. I believe this simile largely speaks the truth. The green and budding twigs may represent existing species; and those produced during each former year may represent the long succession of extinct species. At each period of growth all the growing twigs have tried to branch out on all sides, and to overtop and kill the surrounding twigs and branches, in the same manner as species and groups of species have tried to overmaster other species in the great battle for life. The limbs divided into great branches, and these into lesser and lesser branches, were themselves once, when the tree was small, budding twigs; and this connexion of the former and present buds by ramifying branches may well represent the classification of all extinct and living species in groups subordinate to groups. Of the many twigs which flourished when the tree was a mere bush, only two or three, now grown into great branches, yet survive and bear all the other branches; so with the species which lived during long-past geological periods, very few now have living and modified descendants. From the first growth of the tree, many a limb and branch has decayed and dropped off; and these lost branches of various sizes may represent those whole orders, families, and genera which have now no living representatives, and which are known to us only from having been found in a fossil state. As we here and there see a thin straggling branch springing from a fork low down in a tree, and which by some chance has been favoured and is still alive on its summit, so we occasionally see an animal like the Ornithorhynchus or Lepidosiren, which in some small degree connects by its affinities two large branches of life, and which has apparently been saved from fatal competition by having inhabited a protected station. As buds give rise by growth to fresh buds, and these, if vigorous, branch out and overtop on all sides many a feebler branch, so by generation I believe it has been with the great Tree of Life, which fills with its dead and broken branches the crust of the earth, and covers the surface with its ever branching and beautiful ramifications.

Editor's Introduction to Chapter VI

In this chapter Darwin will raise what he understands to be the principal difficulties that must be faced by the theory of Natural Selection. He begins the discussion by naming four kinds of difficulties. One of them, sterility of hybrids between species, is no longer considered the problem Darwin thought it was; but three difficulties—gaps in the fossil record, complex structures, and elaborate instincts—continue to be invoked today by Darwin's critics, whether speaking from a scientific or a religious standpoint. Do not be surprised if you find that Darwin does not really settle these issues, for a little reflection will show that they probably cannot be definitively answered. The most Darwin can do is to show that they are not *in principle* recalcitrant to explanation by natural selection.

Several considerations limit the explanatory ability of Darwin's theory. First, and most important, the theory is concerned with phenomena that are historical in character, and therefore not fully present to us. The origins of the vertebrate eye, for example, are buried in the past; and since, for the most part, only the hard parts of animals and plants are preserved in the fossil record, there can be practically no material evidence of the origins of the eye. Moreover, the eyes of modern vertebrates exhibit very little gradation. All vertebrates have eyes that are very similar, making it almost impossible to construct even a theoretical model of the eye's evolution. The situation is different for Arthropods (crustaceans, insects, and spiders—Darwin's Articulata). Among them one finds a great variety of types of eyes, from mere light-sensitive patches to organs that are capable of forming images, detecting motion and distinguishing colors. Such variety offers richer possibilities for constructing an imaginative story, a model, of the evolution of vision among the Arthropods. In doing so, however, it is important to keep in mind that what is being constructed is *only* a likely story. The fossil record can only go so far to support such an account.

At the bottom of page 73 Darwin expresses one of his most important insights into the evolution of complex structures. He points out that in thinking about the evolution of the eye, for example, it very difficult not to view it as a problem of *design*, as an engineering problem like that of devising and producing an optical instrument. Nevertheless, even though we may think we know the best way to design an optical system, there is no reason to assume that the engineering solution has anything to do with how the eye actually evolved. As Darwin puts it, it is presumptuous to assume that the Creator works like a human engineer.

The engineer begins with a plan and uses the best materials available in order to effect that plan. Natural selection, by contrast, takes advantage of and transforms structures that are already present and which may originally have had a very different function. The great modern biologist François Jacob once said that an appropriate image for evolution is not the work of an engineer, but rather the activity of a handyman who utilizes whatever materials are available and who adapts his plans to particular circumstances. But even this image is misleading, for it retains the element of *intention.* It is often difficult to keep in mind that the evolutionary process which produced the vertebrate eye was not aiming at the eye in any way. It was not, for Darwin, aiming at anything. You can see this in Darwin's account of the transformation of the swimbladder of fishes into a lung (see page 75); there he envisions the unguided change of a structure that originally served to facilitate flotation into one that became useful for breathing air. Notice, in this example, how far Darwin has stretched his earlier analogy of the skilled breeder!

The engineering paradigm is not the only alternative to natural selection. Darwin is aware of another possibility—that complex structures and habits arose not gradually (as would be the case if brought about by natural selection alone), but suddenly, constituting an abrupt and large-scale change in the organism. Darwin knows that such a thing is not impossible in a world that is governed by the struggle for life. However, as he says, it is the antithesis of his theory of slow incremental change. Since small variations in structure and habit are observed *frequently* in animals and plants, he founds his theory upon such incremental variations, rather than on large changes which are rare—and which, moreover, when they do occur are almost always detrimental to the organism. Thus, while Darwin acknowledges how difficult it is to explain the origin of complex structures by natural selection, he sees clearly that the alternative accounts are even less satisfactory.

It is interesting to note that the three major difficulties mentioned by Darwin hang together. If large-scale changes happen suddenly rather than gradually, then the absence of transitional forms in the fossil record is not due to the record's incompleteness, but rather to the fact that those intermediate forms never existed. As Darwin says, if even one case could be made that some organ or instinct could not have arisen gradually, then his theory of slow and gradual change would be seriously weakened. For that reason much rides on this chapter, and the most Darwin can do is show that his point of view is not the least probable one.

The issues Darwin is raising are not merely of historical interest but are present today in the thinking of evolutionary biologists. There are contemporary biologists who believe that the most prodigious changes in the past—those that represent the differences between the main types of animals and plants (between Arthropods and Vertebrates, for example)—did not occur gradually but arose with relative suddenness. One of the best known proponents of that view is the late Stephen J. Gould, who put forward the theory of "punctuated equilibria." Gould argued that large-scale evolutionary changes are limited to short periods of time, interspersed with very long intervals during which not very much happens. Gould did not deny natural selection, but he gave to it a minor role, that of weeding out maladapted types, rather than the essential role with which Darwin credited it—that of actually shaping the organism gradually over time. The ongoing vitality of such discussions illustrates that Darwin is to be appreciated, not for having established the complete theoretical structure of evolutionary biology, but rather for having initiated a certain way of thinking about nature—a way that is constantly being refined and even revolutionized. This is a more modest, and a more realistic, way of thinking about the founding of any science.

CHAPTER VI

DIFFICULTIES ON THEORY

LONG before having arrived at this part of my work, a crowd of difficulties will have occurred to the reader. Some of them are so grave that to this day I can never reflect on them without being staggered; but, to the best of my judgment, the greater number are only apparent, and those that are real are not, I think, fatal to my theory.

These difficulties and objections may be classed under the following heads:—Firstly, why, if species have descended from other species by insensibly fine gradations, do we not everywhere see innumerable transitional forms? Why is not all nature in confusion instead of the species being, as we see them, well defined?

Secondly, is it possible that an animal having, for instance, the structure and habits of a bat, could have been formed by the modification of some animal with wholly different habits? Can we believe that natural selection could produce, on the one hand, organs of trifling importance, such as the tail of a giraffe, which serves as a fly-flapper, and, on the other hand, organs of such wonderful structure, as the eye, of which we hardly as yet fully understand the inimitable perfection?

Thirdly, can instincts be acquired and modified through natural selection? What shall we say to so marvellous an instinct as that which leads the bee to make cells, which have practically anticipated the discoveries of profound mathematicians? Fourthly, how can we account for species, when crossed, being sterile and producing sterile offspring, whereas, when varieties are crossed, their fertility is unimpaired?

The two first heads shall be here discussed—Instinct and Hybridism in separate chapters.

On the absence or rarity of transitional varieties. As natural selection acts solely by the preservation of profitable modifications, each new form will tend in a fully-stocked country to take the place of, and finally to exterminate, its own less improved parent or other less-favoured forms with which it comes into competition. Thus extinction and natural selection will, as we have seen, go hand in hand. Hence, if we look at each species as descended from some other unknown form, both the parent and all the transitional varieties will generally have been exterminated by the very process of formation and perfection of the new form.

But, as by this theory innumerable transitional forms must have existed, why do we not find them embedded in countless numbers in the crust of the earth? It will be much more convenient to discuss this question in the chapter on the Imperfection of the geological record; and I will here only state that I believe the answer mainly lies in the record being incomparably less perfect than is generally supposed; the imperfection of the record being chiefly due to organic beings not inhabiting profound depths of the sea, and to their remains being embedded and preserved to a future age only in masses of sediment sufficiently thick and extensive to withstand an enormous amount of future degradation; and such fossiliferous masses can be accumulated only where much sediment is deposited on the shallow bed of the sea, whilst it slowly subsides. These contingencies will concur only rarely, and after enormously long intervals. Whilst the bed of the sea is stationary or is rising, or when very little sediment is being deposited, there will be blanks in our geological history. The crust of the earth is a vast museum; but the natural collections have been made only at intervals of time immensely remote.

* * *

Organs of extreme perfection and complication. To suppose that the eye, with all its inimitable contrivances for adjusting the focus to different distances, for admitting different amounts of light, and for the correction of spherical and chromatic aberration, could have been formed by natural selection, seems, I freely confess, absurd in the highest possible degree. Yet reason tells me, that if numerous gradations from a perfect and complex eye to one very imperfect and simple, each grade being useful to its possessor, can be shown to exist; if further, the eye does vary ever so slightly, and the variations be inherited, which is certainly the case; and if any variation or modification in the organ be ever useful to an animal under changing conditions of life, then the difficulty of believing that a perfect and complex eye could be formed by natural selection, though insuperable by our imagination, can hardly be considered real. How a nerve comes to be sensitive to light, hardly concerns us more than how life itself first originated; but I may remark that several facts make me suspect that any sensitive nerve may be rendered sensitive to light, and likewise to those coarser vibrations of the air which produce sound.

In looking for the gradations by which an organ in any species has been perfected, we ought to look exclusively to its lineal ancestors; but this is scarcely ever possible, and we are forced in each case to look to species of the same group, that is to the collateral descendants from the same original parent-form, in order to see what gradations are possible, and for the chance of some gradations having been transmitted from the earlier stages of descent, in an unaltered or little altered condition. Amongst existing Vertebrata, we find but a small amount of gradation in the structure of the eye, and from fossil species we can learn nothing on this head. In this great class we should probably have to descend far beneath the lowest known fossiliferous stratum to discover the earlier stages, by which the eye has been perfected.

In the Articulata[16] we can commence a series with an optic nerve merely coated with pigment, and without any other mechanism; and from this low stage, numerous gradations of structure, branching off in two fundamentally different lines, can be shown to exist, until we reach a moderately high stage of perfection. In certain crustaceans,[17] for instance, there is a double cornea, the inner one divided into facets, within each of which there is a lens shaped swelling. In other crustaceans the transparent cones which are coated by pigment, and which properly act only by excluding lateral pencils of light, are convex at their upper ends and must act by convergence; and at their lower ends there seems to be an imperfect vitreous substance. With these facts, here far too briefly and imperfectly given, which show that there is much graduated diversity in the eyes of living crustaceans, and bearing in mind how small the number of living animals is in proportion to those which have become extinct, I can see no very great difficulty (not more than in the case of many other structures) in believing that natural selection has converted the simple apparatus of an optic nerve merely coated with pigment and invested by transparent membrane, into an optical instrument as perfect as is possessed by any member of the great Articulate class.

He who will go thus far, if he find on finishing this treatise that large bodies of facts, otherwise inexplicable, can be explained by the theory of descent, ought not to hesitate to go further, and to admit that a structure even as perfect as the eye of an eagle might be formed by natural selection, although in this case he does not know any of the transitional grades. His reason ought to conquer his imagination; though I have felt the difficulty far too keenly to be surprised at any degree of hesitation in extending the principle of natural selection to such startling lengths.

It is scarcely possible to avoid comparing the eye to a telescope. We know that this instrument has been perfected by the long-continued efforts of the highest human intellects; and we naturally infer that the eye has been formed by a somewhat analogous process. But may not this inference be presumptuous? Have we any right to assume that the Creator works

16. *Articulata*: a class of animals whose body surface is divided into segments having jointed legs; the class includes insects and crustaceans.

17. *crustaceans*: within the Articulata, a class of aquatic animals having a hard shell and numerous legs; examples include crabs, lobsters, and shrimp.

by intellectual powers like those of man? If we must compare the eye to an optical instrument, we ought in imagination to take a thick layer of transparent tissue, with a nerve sensitive to light beneath, and then suppose every part of this layer to be continually changing slowly in density, so as to separate into layers of different densities and thicknesses, placed at different distances from each other, and with the surfaces of each layer slowly changing in form. Further we must suppose that there is a power always intently watching each slight accidental alteration in the transparent layers; and carefully selecting each alteration which, under varied circumstances, may in any way, or in any degree, tend to produce a distincter image. We must suppose each new state of the instrument to be multiplied by the million; and each to be preserved till a better be produced, and then the old ones to be destroyed. In living bodies, variation will cause the slight alterations, generation will multiply them almost infinitely, and natural selection will pick out with unerring skill each improvement. Let this process go on for millions on millions of years; and during each year on millions of individuals of many kinds; and may we not believe that a living optical instrument might thus be formed as superior to one of glass, as the works of the Creator are to those of man?

If it could be demonstrated that any complex organ existed, which could not possibly have been formed by numerous, successive, slight modifications, my theory would absolutely break down. But I can find out no such case. No doubt many organs exist of which we do not know the transitional grades, more especially if we look to much-isolated species, round which, according to my theory, there has been much extinction. Or again, if we look to an organ common to all the members of a large class, for in this latter case the organ must have been first formed at an extremely remote period, since which all the many members of the class have been developed; and in order to discover the early transitional grades through which the organ has passed, we should have to look to very ancient ancestral forms, long since become extinct.

We should be extremely cautious in concluding that an organ could not have been formed by transitional gradations of some kind. Numerous cases could be given amongst the lower animals of the same organ performing at the same time wholly distinct functions; thus the alimentary canal respires, digests, and excretes in the larva of the dragon-fly and in the fish Cobites. In the Hydra, the animal may be turned inside out, and the exterior surface will then digest and the stomach respire. In such cases natural selection might easily specialise, if any advantage were thus gained, a part or organ, which had performed two functions, for one function alone, and thus wholly change its nature by insensible steps. Two distinct organs sometimes perform simultaneously the same function in the same individual; to give one instance, there are fish with gills or branchiae[18] that breathe the air dissolved in the water, at the same time that they breathe free air in their swimbladders, this latter organ having a ductus pneumaticus for its supply, and being divided by highly vascular partitions. In these cases, one of the two organs might with ease be modified and perfected so as to perform all the work by itself, being aided during the process of modification by the other organ; and then this other organ might be modified for some other and quite distinct purpose, or be quite obliterated.

The illustration of the swimbladder in fishes is a good one, because it shows us clearly the highly important fact that an organ originally constructed for one purpose, namely flotation, may be converted into one for a wholly different purpose, namely respiration. The swimbladder has, also, been worked in as an accessory to the auditory organs of certain fish, or, for I do not know which view is now generally held, a part of the auditory apparatus has been worked in as a complement to the swimbladder. All physiologists admit that the swimbladder is homologous, or "ideally similar," in position and structure with the lungs of the higher vertebrate animals: hence there seems to me to be no great difficulty in believing that natural selection has actually converted a swimbladder into a lung, or organ used exclusively for respiration.

I can, indeed, hardly doubt that all vertebrate animals having true lungs have descended by ordinary generation from an ancient prototype, of which we know nothing, furnished with a floating apparatus or swimbladder. We can thus, as I infer from Professor Owen's interesting description of these parts, understand the strange fact that every particle of food and drink which we swallow has to pass over the orifice of the trachea, with some risk of falling into the lungs, notwithstanding the beautiful contrivance by which the glottis is closed. In the higher Vertebrata the branchiae have wholly disappeared—the slits on the sides of the neck and the loop-like course of the arteries still marking in the embryo their former position. But it is conceivable that the now utterly lost branchiae might have been gradually worked in by natural selection for some quite distinct purpose: in the same manner as, on the view entertained by some naturalists that the branchiae and dorsal scales of Annelids are homologous with the wings and wing-covers of insects, it is probable that organs which at a very ancient period served for respiration have been actually converted into organs of flight.

* * *

18. *gills or branchiae*: Darwin is using these terms synonymously; but *branchiae*, deriving from a Greek word for "fin," can sometimes have a wider application, referring to an organ's shape rather than to the specific function of extracting oxygen from water.

Although we must be extremely cautious in concluding that any organ could not possibly have been produced by successive transitional gradations, yet, undoubtedly, grave cases of difficulty occur, some of which will be discussed in my future work.

One of the gravest is that of neuter insects, which are often very differently constructed from either the males or fertile females; but this case will be treated of in the next chapter. The electric organs of fishes offer another case of special difficulty; it is impossible to conceive by what steps these wondrous organs have been produced; but, as Owen and others have remarked, their intimate structure closely resembles that of common muscle; and as it has lately been shown that Rays have an organ closely analogous to the electric apparatus, and yet do not, as Matteuchi asserts, discharge any electricity, we must own that we are far too ignorant to argue that no transition of any kind is possible.

The electric organs offer another and even more serious difficulty; for they occur in only about a dozen fishes, of which several are widely remote in their affinities. Generally when the same organ appears in several members of the same class, especially if in members having very different habits of life, we may attribute its presence to inheritance from a common ancestor; and its absence in some of the members to its loss through disuse or natural selection. But if the electric organs had been inherited from one ancient progenitor thus provided, we might have expected that all electric fishes would have been specially related to each other. Nor does geology at all lead to the belief that formerly most fishes had electric organs, which most of their modified descendants have lost. The presence of luminous organs in a few insects, belonging to different families and orders, offers a parallel case of difficulty. Other cases could be given; for instance in plants, the very curious contrivance of a mass of pollen-grains, borne on a foot-stalk with a sticky gland at the end, is the same in Orchis and Asclepias,—genera almost as remote as possible amongst flowering plants. In all these cases of two very distinct species furnished with apparently the same anomalous organ, it should be observed that, although the general appearance and function of the organ may be the same, yet some fundamental difference can generally be detected. I am inclined to believe that in nearly the same way as two men have sometimes independently hit on the very same invention, so natural selection, working for the good of each being and taking advantage of analogous variations, has sometimes modified in very nearly the same manner two parts in two organic beings, which owe but little of their structure in common to inheritance from the same ancestor.

Although in many cases it is most difficult to conjecture by what transitions an organ could have arrived at its present state; yet, considering that the proportion of living and known forms to the extinct and unknown is very small, I have been astonished how rarely an organ can be named, towards which no transitional grade is known to lead. The truth of this remark is indeed shown by that old canon in natural history of "Natura non facit saltum."[19] We meet with this admission in the writings of almost every experienced naturalist; or, as Milne Edwards has well expressed it, nature is prodigal in variety, but niggard in innovation. Why, on the theory of Creation, should this be so? Why should all the parts and organs of many independent beings, each supposed to have been separately created for its proper place in nature, be so invariably linked together by graduated steps? Why should not Nature have taken a leap from structure to structure? On the theory of natural selection, we can clearly understand why she should not; for natural selection can act only by taking advantage of slight successive variations; she can never take a leap, but must advance by the shortest and slowest steps.

* * *

19. *Natura non facit saltum*: Nature does not make a leap. This "old canon in natural history" had been invoked also in physics and law; but Darwin is probably citing Linnaeus's *Philosophia Botanica* (1751), where, however, the plural *saltus*, not the singular *saltum*, is used.

Editor's Introduction to Chapter XI

Darwin thought that the distribution of animals and plants around the globe was one of the most powerful pieces of evidence in favor of his historical approach to the origin of the variety of life on earth. It was also the source of one of his most perplexing problems. That problem was solved, and the explanatory power of Darwin's theory revealed, by his account of the influence of the Pleistocene Ice Ages on the contemporary distribution of animals and plants. But before we can appreciate his reasoning we must familiarize ourselves with some facts, some concepts, and some terminology.

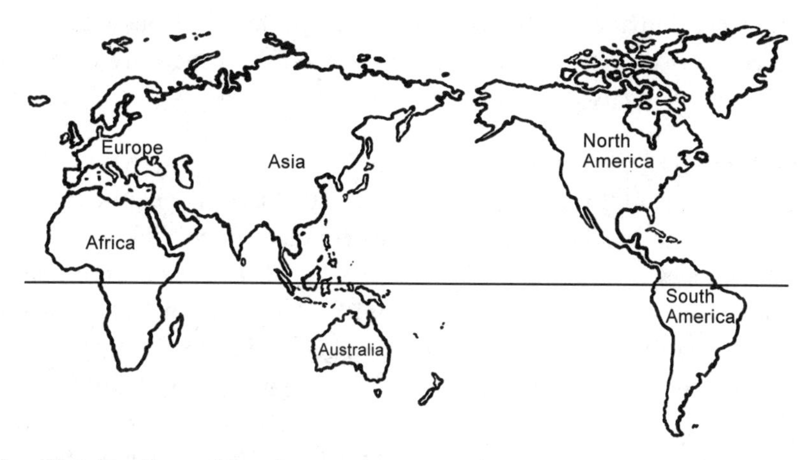

The distribution of land masses

Above is a map of the nonpolar regions of the earth. Notice that the land masses are not evenly distributed, but are largely concentrated in the northern hemisphere; in fact, more than 70% of dry land lies above the equator. Notice also that these masses taper to points towards the south—the Cape of Good Hope in Africa, the tip of India, and Cape Horn in South America. Finally, observe that the only large independent land mass in the nonpolar southern hemisphere is Australia. (Antarctica did not figure into Darwin's account because it was so poorly known.) These facts are crucial to Darwin's argument that the evolution of animals and plants owes more to the pattern of land masses than it does to the climatic conditions in their respective areas.

Through the relations of land masses, some populations are brought together while other populations are isolated from one another. One of Darwin's favorite examples is the dramatic difference between animals and plants in the tropics of the Americas, on the one hand, and those of Africa and Southeast Asia, on the other. The climatic conditions in those areas are very much the same, hot and humid; but the areas are separated from one another by oceans and long distances, so that each area has its own characteristic fauna and flora. For a simple example, compare jaguars and pumas in South America, lions and leopards in Africa, and tigers and leopards in Southeast Asia. These animals are all large feline predators, but they differ greatly, and their differences depend on their relative geographical isolation. By contrast, the brown bear and the gray wolf are the main large predators throughout the entire span of Eurasia and North America. Eurasia and North America are no more climactically similar to one another than are the various tropical regions, but they are contiguous, or almost contiguous, with one another. It is Darwin's contention that such patterns of distribution can only be explained by selection, not by direct adaptation to particular climatic conditions. The theory of direct adaptation, without selection, was that of Lamarck (1744–1829); it was the only competing theory that, like Darwin's, was historical in character.

Single centers of creation

Some points in this chapter bear on an issue that is not explicitly raised by Darwin, but which is the basis of his concern. According to Darwin's theory, whenever one species evolves into another, a very large number of circumstances—including variations in the organism and changing environmental conditions—must come together in the right order. The number of coincident factors required is so great, in fact, that the chance of identical series of changes happening more than once is virtually negligible. A species is thus a unique production of nature happening in a localized region of space and time and which, if it vanishes, will not occur again.

This uniqueness of the species creates a problem for Darwin, for he is obliged, by the foregoing reasoning, to refer an entire species to a single, original, population. This requirement, that each species is ultimately to be traced to a specific, localized, original population, constitutes Darwin's doctrine of "single centers of creation." But if that doctrine is correct, how can it be reconciled with the fact that we often find one and the same species distributed in a number of populations that are both geographically separate and very distant from one another?

In cases like these, Darwin has to explain how a single original population can manage to spread out to isolated regions. That problem would be solved if he can find evidence that the geographical isolation which exists today did not exist in the past, but arose more recently. In this chapter Darwin thinks he has found one major instance in which an apparently impossible problem of distribution is solved through evidence that the earth's climate was once very different from what we experience now.

Geological ages

Geologists have devised a system of names for past periods of time that are characterized by their different assemblages of life forms. The Mesozoic Era, for example, when dinosaurs flourished, extended from 248 million years ago to 65 million years ago. The most recent era is the Cenozoic, which began about 65 million years ago and is characterized by the rise of mammals and flowering plants. That era, in turn, is subdivided into a number of epochs. Darwin focuses on the two most recent epochs—the Pliocene, which began about 5 million years ago, followed by the Pleistocene, which began about 2 million years ago. The Pleistocene is known as a period of repeated *glaciations*, or ice ages. We live now a bit after the end of the last ice age.

Ice ages

The existence of ice ages began to be suspected in the late 18th century, but it was less than 20 years prior to publication of the *Origin* that both their existence and their extent became accepted by the scientific community. The most interesting work had been done by the great Swiss geologist and biologist, Louis Agassiz, who in 1837 gave a celebrated lecture and in 1840 published a book describing the work he had done in the Swiss Alps, England, and Scotland, showing that in the not too distant past the whole of northern Europe had been covered by a great sheet of ice a mile or more thick. These publications had a great impact on geologists. Soon after, it was shown that North America had also been covered by a great mile-thick ice sheet.

We are perhaps so accustomed to the idea of ice ages, it may be difficult to imagine what a radical idea it was when originally proposed, for it implied that in the not-too-distant past the face of the earth was fundamentally different from what it is today. In the mid-1800s there were no contemporary examples of continental glaciations, since the extent of the Greenland and Antarctic ice sheets had not yet been discovered. The only glaciers known were the relatively small ones in mountain ranges. Darwin himself was very resistant to the idea of ice ages for quite a long time, and he wrote extensively in opposition to Agassiz; finally, however, Darwin became persuaded. His initial resistance came, at least in part, from an aversion he shared with Lyell to claims that the earth had ever been subject to large-scale unique changes. It is indeed ironic that an idea which Darwin so long resisted became at last the centerpiece of one of the most persuasive arguments for his theory.

The argument

We are now in a position to study and appreciate Darwin's argument in this chapter. Here are a few suggestions for approaching it. In the first place, notice that it takes the form of a narrative, a story, of how the peculiar distribution of animals and plants in the mountains of Europe came about. The story shows how this pattern of distribution can be explained by inheritance from ancestors who migrated under the influence of the advance and retreat of glaciers; and that it cannot be reasonably explained as the result either of the direct effect of the environment on animals and plants, nor as the result of a design. Bear in mind that the tops of mountains are regions which are isolated from one another just as much as are lakes and oceanic islands. However, the advance and retreat of glaciers makes their isolation more variable than that of other islands. It is this variability that makes them interesting. As you read Darwin's account, try to trace the role that natural selection plays in the argument.

CHAPTER XI

GEOGRAPHICAL DISTRIBUTION

IN considering the distribution of organic beings over the face of the globe, the first great fact which strikes us is, that neither the similarity nor the dissimilarity of the inhabitants of various regions can be accounted for by their climatal and other physical conditions. Of late, almost every author who has studied the subject has come to this conclusion. The case of America alone would almost suffice to prove its truth: for if we exclude the northern parts where the circumpolar land is almost continuous, all authors agree that one of the most fundamental divisions in geographical distribution is that between the New and Old Worlds; yet if we travel over the vast American continent, from the central parts of the United States to its extreme southern point, we meet with the most diversified conditions; the most humid districts, arid deserts, lofty mountains, grassy plains, forests, marshes, lakes, and great rivers, under almost every temperature. There is hardly a climate or condition in the Old World which cannot be paralleled in the New—at least as closely as the same species generally require; for it is a most rare case to find a group of organisms confined to any small spot, having conditions peculiar in only a slight degree; for instance, small areas in the Old World could be pointed out hotter than any in the New World, yet these are not inhabited by a peculiar fauna or flora.[20] Notwithstanding this parallelism in the conditions of the Old and New Worlds, how widely different are their living productions!

In the southern hemisphere, if we compare large tracts of land in Australia, South Africa, and western South America, between latitudes 25° and 35°, we shall find parts extremely similar in all their conditions, yet it would not be possible to point out three faunas and floras more utterly dissimilar. Or again we may compare the productions of South America south of lat. 35° with those north of 25°, which consequently inhabit a considerably different climate, and they will be found incomparably more closely related to each other, than they are to the productions of Australia or Africa under nearly the same climate. Analogous facts could be given with respect to the inhabitants of the sea.

20. fauna or flora: The collective stock of animals or plants, respectively, that populate a particular region or geological era.

A second great fact which strikes us in our general review is, that barriers of any kind, or obstacles to free migration, are related in a close and important manner to the differences between the productions of various regions. We see this in the great difference of nearly all the terrestrial productions of the New and Old Worlds, excepting in the northern parts, where the land almost joins, and where, under a slightly different climate, there might have been free migration for the northern temperate forms, as there now is for the strictly arctic productions. We see the same fact in the great difference between the inhabitants of Australia, Africa, and South America under the same latitude: for these countries are almost as much isolated from each other as is possible. On each continent, also, we see the same fact; for on the opposite sides of lofty and continuous mountain-ranges, and of great deserts, and sometimes even of large rivers, we find different productions; though as mountain-chains, deserts, &c., are not as impassable, or likely to have endured so long as the oceans separating continents, the differences are very inferior in degree to those characteristic of distinct continents.

* * *

A third great fact, partly included in the foregoing statements, is the affinity of the productions of the same continent or sea, though the species themselves are distinct at different points and stations. It is a law of the widest generality, and every continent offers innumerable instances. Nevertheless the naturalist in travelling, for instance, from north to south never fails to be struck by the manner in which successive groups of beings, specifically distinct, yet clearly related, replace each other. He hears from closely allied, yet distinct kinds of birds, notes nearly similar, and sees their nests similarly constructed, but not quite alike, with eggs coloured in nearly the same manner. The plains near the Straits of Magellan are inhabited by one species of Rhea (American ostrich), and northward the plains of La Plata by another species of the same genus; and not by a true ostrich or emeu, like those found in Africa and Australia under the same latitude. On these same plains of la Plata, we see the agouti and bizcacha, animals having nearly the same habits as our hares and rabbits and belonging to the same order of Rodents, but they plainly display an American type of structure. We ascend the lofty peaks of the Cordillera and we find an alpine species of bizcacha; we look to the waters, and we do not find the beaver or musk-rat, but the coypu and capybara, rodents of the American type. Innumerable other instances could be given. If we look to the islands off the American shore, however much they may differ in geological structure, the inhabitants, though they may be all peculiar species, are essentially

American. We may look back to past ages, as shown in the last chapter,[21] and we find American types then prevalent on the American continent and in the American seas. We see in these facts some deep organic bond, prevailing throughout space and time, over the same areas of land and water, and independent of their physical conditions. The naturalist must feel little curiosity, who is not led to inquire what this bond is.

This bond, on my theory, is simply inheritance, that cause which alone, as far as we positively know, produces organisms quite like, or, as we see in the case of varieties nearly like each other. The dissimilarity of the inhabitants of different regions may be attributed to modification through natural selection, and in a quite subordinate degree to the direct influence of different physical conditions. The degree of dissimilarity will depend on the migration of the more dominant forms of life from one region into another having been effected with more or less ease, at periods more or less remote;—on the nature and number of the former immigrants;—and on their action and reaction, in their mutual struggles for life;—the relation of organism to organism being, as I have already often remarked, the most important of all relations. Thus the high importance of barriers comes into play by checking migration; as does time for the slow process of modification through natural selection. Widely-ranging species, abounding in individuals, which have already triumphed over many competitors in their own widely-extended homes will have the best chance of seizing on new places, when they spread into new countries. In their new homes they will be exposed to new conditions, and will frequently undergo further modification and improvement; and thus they will become still further victorious, and will produce groups of modified descendants. On this principle of inheritance with modification, we can understand how it is that sections of genera, whole genera, and even families are confined to the same areas, as is so commonly and notoriously the case.

I believe, as was remarked in the last chapter, in no law of necessary development. As the variability of each species is an independent property, and will be taken advantage of by natural selection, only so far as it profits the individual in its complex struggle for life, so the degree of modification in different species will be no uniform quantity. If, for instance, a number of species, which stand in direct competition with each other, migrate in a body into a new and afterwards isolated country, they will be little liable to modification; for neither migration nor isolation in themselves can do anything. These principles come into play only by bringing organisms into new relations with each other, and in a lesser degree with the surrounding

21. *the last chapter*: Here and below, Darwin's Chapter X, not included in our selections.

physical conditions. As we have seen in the last chapter that some forms have retained nearly the same character from an enormously remote geological period, so certain species have migrated over vast spaces, and have not become greatly modified.

* * *

We are thus brought to the question which has been largely discussed by naturalists, namely, whether species have been created at one or more points of the earth's surface. Undoubtedly there are very many cases of extreme difficulty, in understanding how the same species could possibly have migrated from some one point to the several distant and isolated points, where now found. Nevertheless the simplicity of the view that each species was first produced within a single region captivates the mind. He who rejects it, rejects the *vera causa* of ordinary generation with subsequent migration, and calls in the agency of a miracle. It is universally admitted, that in most cases the area inhabited by a species is continuous; and when a plant or animal inhabits two points so distant from each other, or with an interval of such a nature, that the space could not be easily passed over by migration, the fact is given as something remarkable and exceptional. The capacity of migrating across the sea is more distinctly limited in terrestrial mammals, than perhaps in any other organic beings; and, accordingly, we find no inexplicable cases of the same mammal inhabiting distant points of the world. No geologist will feel any difficulty in such cases as Great Britain having been formerly united to Europe, and consequently possessing the same quadrupeds. But if the same species can be produced at two separate points, why do we not find a single mammal common to Europe and Australia or South America? The conditions of life are nearly the same, so that a multitude of European animals and plants have become naturalised in America and Australia; and some of the aboriginal plants are identically the same at these distant points of the northern and southern hemispheres? The answer, as I believe, is, that mammals have not been able to migrate, whereas some plants, from their varied means of dispersal, have migrated across the vast and broken interspace. The great and striking influence which barriers of every kind have had on distribution, is intelligible only on the view that the great majority of species have been produced on one side alone, and have not been able to migrate to the other side. Some few families, many sub-families, very many genera, and a still greater number of sections of genera are confined to a single region; and it has been observed by several naturalists, that the most natural genera, or those genera in which the species are most closely related to each other, are generally local, or confined to one area. What a strange anomaly it would be, if, when coming one step lower in the series, to the individuals of the same species, a directly opposite rule prevailed; and species were not local, but had been produced in two or more distinct areas!

Hence it seems to me, as it has to many other naturalists, that the view of each species having been produced in one area alone, and having subsequently migrated from that area as far as its powers of migration and subsistence under past and present conditions permitted, is the most probable. Undoubtedly many cases occur, in which we cannot explain how the same species could have passed from one point to the other. But the geographical and climatal changes, which have certainly occurred within recent geological times, must have interrupted or rendered discontinuous the formerly continuous range of many species. So that we are reduced to consider whether the exceptions to continuity of range are so numerous and of so grave a nature, that we ought to give up the belief, rendered probable by general considerations, that each species has been produced within one area, and has migrated thence as far as it could. It would be hopelessly tedious to discuss all the exceptional cases of the same species, now living at distant and separated points; nor do I for a moment pretend that any explanation could be offered of many such cases. But after some preliminary remarks, I will discuss a few of the most striking classes of facts; namely, the existence of the same species on the summits of distant mountain-ranges, and at distant points in the arctic and antarctic regions; and secondly (in the following chapter), the wide distribution of freshwater productions; and thirdly, the occurrence of the same terrestrial species on islands and on the mainland, though separated by hundreds of miles of open sea. If the existence of the same species at distant and isolated points of the earth's surface, can in many instances be explained on the view of each species having migrated from a single birthplace; then, considering our ignorance with respect to former climatal and geographical changes and various occasional means of transport, the belief that this has been the universal law, seems to me incomparably the safest.

* * *

Before discussing the three classes of facts, which I have selected as presenting the greatest amount of difficulty on the theory of "single centres of creation," I must say a few words on the means of dispersal.

* * *

Dispersal during the Glacial period. The identity of many plants and animals, on mountain-summits, separated from each other by hundreds of miles of lowlands, where the Alpine species could not possibly exist, is one of the most striking cases known of the same species living at distant points, without the apparent possibility of their having migrated from one to the other. It is indeed a remarkable fact to see so many of the same plants living on the snowy regions of the Alps or Pyrenees, and in the extreme northern parts of Europe; but it is far more remarkable, that the plants on the White Mountains, in the United States of America, are all the same with those of Labrador, and nearly all the same, as we hear from Asa Gray, with those on the loftiest mountains of Europe. Even as long ago as 1747, such facts led Gmelin to conclude that the same species must have been independently created at several distinct points; and we might hve remained in this same belief, had not Agassiz and others called vivid attention to the Glacial period,[22] which, as we shall immediately see, affords a simple explanation of these facts. We have evidence of almost every conceivable kind, organic and inorganic, that within a very recent geological period, central Europe and North America suffered under an Arctic climate. The ruins of a house burnt by fire do not tell their tale more plainly, than do the mountains of Scotland and Wales, with their scored flanks, polished surfaces, and perched boulders, of the icy streams with which their valleys were lately filled. So greatly has the climate of Europe changed, that in Northern Italy, gigantic moraines,[23] left by old glaciers, are now clothed by the vine and maize. Throughout a large part of the United States, erratic boulders, and rocks scored by drifted icebergs and coast-ice, plainly reveal a former cold period.

The former influence of the glacial climate on the distribution of the inhabitants of Europe, as explained with remarkable clearness by Edward Forbes, is substantially as follows. But we shall follow the changes more readily, by supposing a new glacial period to come slowly on, and then pass away, as formerly occurred. As the cold came on, and as each more southern zone became fitted for arctic beings and ill-fitted for their former more temperate inhabitants, the latter would be supplanted and arctic productions would take their places. The inhabitants of the more temperate regions would at the same time travel southward, unless they were stopped by barriers, in which case they would perish. The mountains would become

22. *Glacial period*: The editor's introduction to this chapter described Louis Agassiz's theory that in the comparatively recent geologic past, vast ice sheets of ice had covered much of Europe and North America. Darwin speaks only of a single such glacial period, but by the end of the 19th century, evidence for multiple glacial cycles began to appear. With the development of radiometric dating in the early 20th century, it became possible to correlate geological events with calender years; the most recent glacial period is now thought to have ended about 10,000 years ago.

23. *moraine*: an area covered by rocks and debris transported, and subsequently deposited by, a glacier.

covered with snow and ice, and their former Alpine inhabitants would descend to the plains. By the time that the cold had reached its maximum, we should have a uniform arctic fauna and flora, covering the central parts of Europe, as far south as the Alps and Pyrenees, and even stretching into Spain. The now temperate regions of the United States would likewise be covered by arctic plants and animals, and these would be nearly the same with those of Europe; for the present circumpolar inhabitants, which we suppose to have everywhere travelled southward, are remarkably uniform round the world. We may suppose that the Glacial period came on a little earlier or later in North America than in Europe, so will the southern migration there have been a little earlier or later; but this will make no difference in the final result.

As the warmth returned, the arctic forms would retreat northward, closely followed up in their retreat by the productions of the more temperate regions. And as the snow melted from the bases of the mountains, the arctic forms would seize on the cleared and thawed ground, always ascending higher and higher, as the warmth increased, whilst their brethren were pursuing their northern journey. Hence, when the warmth had fully returned, the same arctic species, which had lately lived in a body together on the lowlands of the Old and New Worlds, would be left isolated on distant mountain-summits (having been exterminated on all lesser heights) and in the arctic regions of both hemispheres.

Thus we can understand the identity of many plants at points so immensely remote as on the mountains of the United States and of Europe. We can thus also understand the fact that the Alpine plants of each mountain-range are more especially related to the arctic forms living due north or nearly due north of them: for the migration as the cold came on, and the re-migration on the returning warmth, will generally have been due south and north. The Alpine plants, for example, of Scotland, as remarked by Mr. H. C. Watson, and those of the Pyrenees, as remarked by Ramond, are more especially allied to the plants of northern Scandinavia; those of the United States to Labrador; those of the mountains of Siberia to the arctic regions of that country. These views, grounded as they are on the perfectly well-ascertained occurrence of a former Glacial period, seem to me to explain in so satisfactory a manner the present distribution of the Alpine and Arctic productions of Europe and America, that when in other regions we find the same species on distant mountain-summits, we may almost conclude without other evidence, that a colder climate permitted their former migration across the low intervening tracts, since become too warm for their existence.

If the climate, since the Glacial period, has ever been in any degree warmer than at present (as some geologists in the United States believe to have been the case, chiefly from the distribution of the fossil Gnathodon), then the arctic and temperate productions will at a very late period have marched a little further north, and subsequently have retreated to their present homes; but I have met with no satisfactory evidence with respect to this intercalated slightly warmer period, since the Glacial period. The arctic forms, during their long southern migration and re-migration northward, will have been exposed to nearly the same climate, and, as is especially to be noticed, they will have kept in a body together; consequently their mutual relations will not have been much disturbed, and, in accordance with the principles inculcated in this volume, they will not have been liable to much modification. But with our Alpine productions, left isolated from the moment of the returning warmth, first at the bases and ultimately on the summits of the mountains, the case will have been somewhat different; for it is not likely that all the same arctic species will have been left on mountain ranges distant from each other, and have survived there ever since; they will, also, in all probability have become mingled with ancient Alpine species, which must have existed on the mountains before the commencement of the Glacial epoch, and which during its coldest period will have been temporarily driven down to the plains; they will, also, have been exposed to somewhat different climatal influences. Their mutual relations will thus have been in some degree disturbed; consequently they will have been liable to modification; and this we find has been the case; for if we compare the present Alpine plants and animals of the several great European mountain-ranges, though very many of the species are identically the same, some present varieties, some are ranked as doubtful forms, and some few are distinct yet closely allied or representative species.

* * *

These cases of relationship, without identity, of the inhabitants of seas now disjoined, and likewise of the past and present inhabitants of the temperate lands of North America and Europe, are inexplicable on the theory of creation. We cannot say that they have been created alike, in correspondence with the nearly similar physical conditions of the areas; for if we compare, for instance, certain parts of South America with the southern continents of the Old World, we see countries closely corresponding in all their physical conditions, but with their inhabitants utterly dissimilar.

But we must return to our more immediate subject, the Glacial period. I am convinced that Forbes's view may be largely extended. In Europe we have the plainest evidence of the cold period, from the western shores of Britain to the Oural range, and southward to the Pyrenees. We may infer, from the frozen mammals and nature of the mountain vegetation, that Siberia was similarly affected. Along the Himalaya, at points 900 miles apart, glaciers have left the marks of their former low descent; and in Sikkim, Dr. Hooker saw maize growing on gigantic ancient moraines.

* * *

On this view of the whole world, or at least of broad longitudinal belts, having been simultaneously colder from pole to pole, much light can be thrown on the present distribution of identical and allied species. In America, Dr. Hooker has shown that between forty and fifty of the flowering plants of Tierra del Fuego, forming no inconsiderable part of its scanty flora, are common to Europe, enormously remote as these two points are; and there are many closely allied species. On the lofty mountains of equatorial America a host of peculiar species belonging to European genera occur. On the highest mountains of Brazil, some few European genera were found by Gardner, which do not exist in the wide intervening hot countries. So on the Silla of Caraccas the illustrious Humboldt long ago found species belonging to genera characteristic of the Cordillera. On the mountains of Abyssinia, several European forms and some few representatives of the peculiar flora of the Cape of Good Hope occur. At the Cape of Good Hope a very few European species, believed not to have been introduced by man, and on the mountains, some few representative European forms are found, which have not been discovered in the intertropical parts of Africa. On the Himalaya, and on the isolated mountain-ranges of the peninsula of India, on the heights of Ceylon, and on the volcanic cones of Java, many plants occur, either identically the same or representing each other, and at the same time representing plants of Europe, not found in the intervening hot lowlands. A list of the genera collected on the loftier peaks of Java raises a picture of a collection made on a hill in Europe! Still more striking is the fact that southern Australian forms are clearly represented by plants growing on the summits of the mountains of Borneo. Some of these Australian forms, as I hear from Dr. Hooker, extend along the heights of the peninsula of Malacca, and are thinly scattered, on the one hand over India and on the other as far as Japan.

On the southern mountains of Australia, Dr. F. Muller has discovered several European species; other species, not introduced by man, occur on the lowlands; and a long list can be given, as I am informed by Dr. Hooker, of European genera, found in Australia, but not in the intermediate torrid regions. In the admirable "Introduction to the Flora of New Zealand," by Dr. Hooker, analogous and striking facts are given in regard to the plants of that large island. Hence we see that throughout the world, the plants growing on the more lofty mountains, and on the temperate lowlands of the northern and southern hemispheres, are sometimes identically the same; but they are much oftener specifically distinct, though related to each other in a most remarkable manner.

This brief abstract applies to plants alone: some strictly analogous facts could be given on the distribution of terrestrial animals. In marine productions, similar cases occur; as an example, I may quote a remark by the highest authority, Prof. Dana, that "it is certainly a wonderful fact that New Zealand should have a closer resemblance in its crustacea to Great Britain, its antipode, than to any other part of the world." Sir J. Richardson, also, speaks of the reappearance on the shores of New Zealand, Tasmania, &c., of northern forms of fish. Dr. Hooker informs me that twenty-five species of Algae are common to New Zealand and to Europe, but have not been found in the intermediate tropical seas.

* * *

Now let us see what light can be thrown on the foregoing facts, on the belief, supported as it is by a large body of geological evidence, that the whole world, or a large part of it, was during the Glacial period simultaneously much colder than at present. The Glacial period, as measured by years, must have been very long; and when we remember over what vast spaces some naturalised plants and animals have spread within a few centuries, this period will have been ample for any amount of migration. As the cold came slowly on, all the tropical plants and other productions will have retreated from both sides towards the equator, followed in the rear by the temperate productions, and these by the arctic; but with the latter we are not now concerned. The tropical plants probably suffered much extinction; how much no one can say; perhaps formerly the tropics supported as many species as we see at the present day crowded together at the Cape of Good Hope, and in parts of temperate Australia. As we know that many tropical plants and animals can withstand a considerable amount of cold, many might have escaped extermination during a moderate fall of temperature, more especially by escaping into the warmest spots. But the great fact to bear in mind is, that all tropical productions will have suffered to a certain extent. On the other hand, the temperate productions, after migrating nearer to the equator, though they will have been placed under somewhat new conditions, will have suffered less. And it is certain that many temperate plants, if protected from the inroads of competitors, can withstand a much warmer climate than their own. Hence, it seems to me possible, bearing in mind that the tropical productions were in a suffering state and could not have presented a firm front against intruders, that a certain number of the more vigorous and dominant temperate forms might have penetrated the native ranks and have reached or even crossed the equator. The invasion would, of course, have been greatly favoured by high land, and perhaps by a dry climate; for Dr. Falconer informs me that it is the damp with the heat of the tropics which is so destructive to perennial plants from a temperate climate. On the other hand, the most humid and hottest districts will have afforded an asylum to the tropical natives. The mountain-ranges north-west of the Himalaya, and the long line of the Cordillera, seem to have afforded two great lines of invasion: and it is a striking fact, lately communicated to me by Dr. Hooker, that all the flowering plants, about forty-six in number, common to Tierra del Fuego and to Europe still exist in North America, which must have lain on the line of march. But I do not doubt that some temperate productions entered and crossed even the *lowlands* of the tropics at the period when the cold was most intense,—when arctic forms had migrated some twenty-five degrees of latitude from their native country and covered the land at the foot of the Pyrenees. At this period of extreme cold, I believe that the climate under the equator at the level of the sea was about the same with that now felt there at the height of six or seven thousand feet. During this the coldest period, I suppose that large spaces of the tropical lowlands were clothed with a mingled tropical and temperate vegetation, like that now growing with strange luxuriance at the base of the Himalaya, as graphically described by Hooker.

Thus, as I believe, a considerable number of plants, a few terrestrial animals, and some marine productions, migrated during the Glacial period from the northern and southern temperate zones into the intertropical regions, and some even crossed the equator. As the warmth returned, these temperate forms would naturally ascend the higher mountains, being exterminated on the lowlands; those which had not reached the equator, would re-migrate northward or southward towards their former homes; but the forms, chiefly northern, which had crossed the equator, would travel still further from their homes into the more temperate latitudes of the opposite hemisphere. Although we have reason to believe from geological evidence that the whole body of arctic shells underwent scarcely any modification during their long southern migration and re-migration northward, the case may have been wholly different with those intruding forms which settled themselves on the intertropical mountains, and in the southern hemisphere. These being surrounded by strangers will have had to compete with many new forms of life; and it is probable that selected modifications in their structure, habits, and constitutions will have profited them. Thus many of these wanderers, though still plainly related by inheritance to their brethren of the northern or southern hemispheres, now exist in their new homes as well-marked varieties or as distinct species.

It is a remarkable fact, strongly insisted on by Hooker in regard to America, and by Alph. de Candolle in regard to Australia, that many more identical plants and allied forms have apparently migrated from the north to the south, than in a reversed direction. We see, however, a few southern vegetable forms on the mountains of Borneo and Abyssinia. I suspect that this preponderant migration from north to south is due to the greater extent of land in the north, and to the northern forms having existed in their own homes in greater numbers, and having consequently been advanced through natural selection and competition to a higher stage of perfection or dominating power, than the southern forms. And thus, when they became commingled during the Glacial period, the northern forms were enabled to beat the less powerful southern forms. Just in the same manner as we see at the present day, that very many European productions cover the ground in La Plata, and in a lesser degree in Australia, and have to a certain extent beaten the natives; whereas extremely few southern forms have become naturalised in any part of Europe, though hides, wool, and other objects likely to carry seeds have been largely imported into Europe during the last two or three centuries from La Plata, and during the last thirty or forty years from Australia. Something of the same kind must have occurred on the intertropical mountains: no doubt before the Glacial period they were stocked with endemic Alpine forms; but these have almost everywhere largely yielded to the more dominant forms, generated in the larger areas and more efficient workshops of the north. In many islands the native productions are nearly equalled or even outnumbered by the naturalised; and if the natives have not been actually exterminated, their numbers have been greatly reduced, and this is the first stage towards extinction. A mountain is an island on the land; and the intertropical mountains before the Glacial period must have been completely isolated; and I believe that the productions of these islands on the land yielded to those produced within the larger areas of the north, just in the same way as the productions of real islands have everywhere lately yielded to continental forms, naturalised by man's agency.

I am far from supposing that all difficulties are removed on the view here given in regard to the range and affinities of the allied species which live in the northern and southern temperate zones and on the mountains of the intertropical regions. Very many difficulties remain to be solved. I do not pretend to indicate the exact lines and means of migration, or the reason why certain species and not others have migrated; why certain species have been modified and have given rise to new groups of forms, and others have remained unaltered. We cannot hope to explain such facts, until we can say why one species and not another becomes naturalised by man's agency in a foreign land; why one ranges twice or thrice as far, and is twice or thrice as common, as another species within their own homes.

* * *

Editor's Introduction to Chapter XIV

We come now to the final chapter of the *Origin*. Darwin titled this chapter "Recapitulation and Conclusion," but we shall pass over Darwin's recapitulation of the argument and turn immediately to his conclusion, which occupies the final third of the chapter. In this discussion Darwin looks ahead to outline how he believes the study of living things will be transformed by his vision of the radical historicity of life

Darwin understands that a historical view of life will be very hard for older naturalists to accept. Traditionally, naturalists have been concerned to identify and characterize the stable forms that underlie the variety of living beings found in the world. They have identified these beings with the species of animals and plants as ordinarily understood—the whitetail deer, the tiger, the earthworm, the red oak, the white pine, and the many other forms of life which we commonly see around us. For these naturalists, all changes in the world of living things were to be understood in relation to those stable and permanent life forms—for if nothing is stable then it is impossible to say that something changes.

Darwin is far from despising this point of view, and to some extent he even shares the aim of identifying the permanent features of the world, the background against which change occurs. But what, for him, *is* permanent? Darwin's argument throughout the book has made it clear that he regards the *geographical race*—a population of living things that share a life in all its aspects, including coming together to produce the next generation—as the fundamental entity. The geographical race is real in the sense that it exists right in front of us—unlike the species, which has to be defined by the scientist. But the geographical race is not static. It can split up or even become extinct. Finally, it is the geographical race that evolves. In our own time the *gene* has been proposed as the fundamental entity. But both the geographical race and the gene are permanent only as historical lineages, not as unchanging kinds of things. Although Darwin does not explicitly discuss the question of permanance, we readers should keep it in mind. The only thing Darwin is sure of is that trying to distinguish the "true" species from mere varieties is futile, for the world is not set up that way.

Darwin also notes the dilemma of those traditional naturalists who have realized that the geological record indicates that the world of living things was very different in the past from the way it is now. Conclusive geological evidence implied the existence of large and imposing animals like dinosaurs, wooly mammoths, and giant sloths. Such animals are not found living anywhere today; and the world was too well-known in the mid-19th century to allow the belief that they were just hidden away in some as-yet-undiscovered regions. Traditional naturalists attempted to reconcile the fossil record with their belief in the fixity of species by appealing to the incompleteness of that record. Rather than interpreting fossils of extinct animals as evidence of prehistoric species that have evolved into other species, they maintained that there had been, in the past, great catastrophes that wiped out all life on earth, and that after each of these the earth was repopulated with a whole new fauna and flora. Thus, although species could not be said to be absolutely fixed and eternal, they were fixed with each new creation.

The great French paleontologist Georges Cuvier (1769–1832) was the foremost proponent of this idea. Cuvier was convinced that the fossil record was best interpreted as showing a succession of living worlds separated from one another by catastrophic destructions. He was, however, completely silent both as to the nature of these catastrophes and to the ways by which the earth was subsequently repopulated. It is this silence that Darwin is criticizing when he writes (page 100): "Although naturalists very properly demand a full explanation of every difficulty from those who believe in the mutability of species, on their own side they ignore the whole subject of the first appearance of species in what they consider reverent silence."

Darwin has no contempt for Cuvier's thinking. In fact Cuvier and his contemporary Lamarck were Darwin's greatest teachers. It should be mentioned that a version of Cuvier's ideas has been recently revived through the discovery of evidence for a number of mass extinctions and subsequent repopulations of the earth in the past.

Pay careful attention to the two paragraphs beginning on the bottom of page 100. For us modern readers, the theory of evolution by natural selection is intimately tied to ideas about the ultimate origin of life on earth. Notice how circumspect Darwin is about this question. Although he is quite sure that the fundamental similarities among living forms imply that life on earth had a single beginning, he is not sure how that beginning happened. In particular, he is not sure how far back in the history of life natural selection was operative. And while he is sure that the differences within a class, like that of the vertebrates, can be accounted for by natural selection, he is not certain that the differences between, say, vertebrates and arthropods are of the same kind. He leaves open the possibility that these large classes may have emerged through principles other than natural selection.

The last few pages of this chapter are among the most lyrical in the whole book; but while they are eminently readable, the profusion of images makes it hard to know exactly what Darwin is saying. Here are a few suggestions for points on which to focus. First, through the image of the Tree of Life, Darwin again invokes genealogy as the basis of the science of classification. He foresees that once naturalists turn to the work of unraveling the real genealogical connections among living things, they will "be freed from the vain search for the undiscovered and undiscoverable essence of the term species" (page 102).

Next, consider Darwin's abrupt and perhaps diffident sentence, "Light will be thrown on the origin of man and his history" (page 104). This is the only place in the book where man himself is said to have a developmental history.

In the last two paragraphs of the chapter, Darwin connects the profusion and beauty of life on earth with *laws*, and even refers to the Newtonian theory of universal gravitation. Perhaps Darwin means to indicate that the science of life has now, like physics, freed itself from dependence on a belief in permanent beings as its basis and is instead, again like physics, to be grounded in laws.

In two critical places, Darwin characterizes life as having been "breathed into" some primorial form or forms: on page 101 when he first mentions the origin of life itself, and again on the final page, as he repeats the thought. The phrase is almost certainly meant to allude to Genesis 2:7, in which God breathed life into the nostrils of the first man. Whether Darwin is actually introducing God into evolutionary theory is harder to say. More likely he is invoking the spirit of the epigraphs at the beginning of the book, which suggest that there need be no conflict between science and religion.

Finally, notice that the last word of the book is "evolved." Darwin uses the word nowhere else in the *Origin*. Originally, the verb "evolve" pointed to the unfolding and becoming present of something that was already existing, but hidden. A flower, for example, could be said to evolve from the bud inasmuch as the flower was already there in the bud and only unfolds itself. But at the end of the book Darwin seems to have given the word a new meaning, one that indicates the coming into being of something that did not previously exist. Darwin's whole argument has directed us to a new and radical understanding of how the forms of life emerged and are still emerging.

CHAPTER XIV

RECAPITULATION AND CONCLUSION

* * *

I have now recapitulated the chief facts and considerations which have thoroughly convinced me that species have changed, and are still slowly changing by the preservation and accumulation of successive slight favourable variations. Why, it may be asked, have all the most eminent living naturalists and geologists rejected this view of the mutability of species? It cannot be asserted that organic beings in a state of nature are subject to no variation; it cannot be proved that the amount of variation in the course of long ages is a limited quantity; no clear distinction has been, or can be, drawn between species and well-marked varieties. It cannot be maintained that species when intercrossed are invariably sterile, and varieties invariably fertile; or that sterility is a special endowment and sign of creation. The belief that species were immutable productions was almost unavoidable as long as the history of the world was thought to be of short duration; and now that we have acquired some idea of the lapse of time, we are too apt to assume, without proof, that the geological record is so perfect that it would have afforded us plain evidence of the mutation of species, if they had undergone mutation.

But the chief cause of our natural unwillingness to admit that one species has given birth to other and distinct species, is that we are always slow in admitting any great change of which we do not see the intermediate steps. The difficulty is the same as that felt by so many geologists, when Lyell first insisted that long lines of inland cliffs had been formed, and great valleys excavated, by the slow action of the coast-waves. The mind cannot possibly grasp the full meaning of the term of a hundred million years; it cannot add up and perceive the full effects of many slight variations, accumulated during an almost infinite number of generations.

Although I am fully convinced of the truth of the views given in this volume under the form of an abstract, I by no means expect to convince experienced naturalists whose minds are stocked with a multitude of facts all viewed, during a long course of years, from a point of view directly opposite to mine. It is so easy to hide our ignorance under such expressions as the "plan of creation," "unity of design," &c., and to think that we give an explanation when we only restate a fact. Any one whose disposition leads him to attach more weight to unexplained difficulties than to the explanation of a certain number of facts will certainly reject my theory. A few naturalists, endowed with much flexibility of mind, and who have already begun to doubt on the immutability of species, may be influenced by this volume; but I look with confidence to the future, to young and rising naturalists, who will be able to view both sides of the question with impartiality. Whoever is led to believe that species are mutable will do good service by conscientiously expressing his conviction; for only thus can the load of prejudice by which this subject is overwhelmed be removed.

Several eminent naturalists have of late published their belief that a multitude of reputed species in each genus are not real species; but that other species are real, that is, have been independently created. This seems to me a strange conclusion to arrive at. They admit that a multitude of forms, which till lately they themselves thought were special creations, and which are still thus looked at by the majority of naturalists, and which consequently have every external characteristic feature of true species,—they admit that these have been produced by variation, but they refuse to extend the same view to other and very slightly different forms. Nevertheless they do not pretend that they can define, or even conjecture, which are the created forms of life, and which are those produced by secondary laws. They admit variation as a *vera causa* in one case, they arbitrarily reject it in another, without assigning any distinction in the two cases. The day will come when this will be given as a curious illustration of the blindness of preconceived opinion. These authors seem no more startled at a miraculous act of creation than at an ordinary birth. But do they really believe that at innumerable periods in the earth's history certain elemental atoms have been commanded suddenly to flash into living tissues? Do they believe that at each supposed act of creation one individual or many were produced? Were all the infinitely numerous kinds of animals and plants created as eggs or seed, or as full grown? and in the case of mammals, were they created bearing the false marks of nourishment from the mother's womb? Although naturalists very properly demand a full explanation of every difficulty from those who believe in the mutability of species, on their own side they ignore the whole subject of the first appearance of species in what they consider reverent silence.

It may be asked how far I extend the doctrine of the modification of species. The question is difficult to answer, because the more distinct the forms are which we may consider, by so much the arguments fall away in force. But some arguments of the greatest weight extend very far. All the members of whole classes can be connected together by chains of affinities, and all can be classified on the same principle, in groups subordinate to groups. Fossil remains sometimes tend to fill up very wide intervals between existing orders. Organs in a rudimentary condition plainly show that an early progenitor had the organ in a fully developed state; and this in some instances necessarily implies an enormous amount of modification in the descendants. Throughout whole classes various structures are formed on the same pattern, and at an embryonic age the species closely resemble each other. Therefore I cannot doubt that the theory of descent with modification embraces all the members of the same class. I believe that animals have descended from at most only four or five progenitors, and plants from an equal or lesser number.

Analogy would lead me one step further, namely, to the belief that all animals and plants have descended from some one prototype. But analogy may be a deceitful guide. Nevertheless all living things have much in common, in their chemical composition, their germinal vesicles, their cellular structure, and their laws of growth and reproduction. We see this even in so trifling a circumstance as that the same poison often similarly affects plants and animals; or that the poison secreted by the gall-fly produces monstrous growths on the wild rose or oak-tree. Therefore I should infer from analogy that probably all the organic beings which have ever lived on this earth have descended from some one primordial form, into which life was first breathed.

When the views entertained in this volume on the origin of species, or when analogous views are generally admitted, we can dimly foresee that there will be a considerable revolution in natural history. Systematists will be able to pursue their labours as at present; but they will not be incessantly haunted by the shadowy doubt whether this or that form be in essence a species. This I feel sure, and I speak after experience, will be no slight relief. The endless disputes whether or not some fifty species of British brambles are true species will cease. Systematists will have only to decide (not that this will be easy) whether any form be sufficiently constant and distinct from other forms, to be capable of definition; and if definable, whether the differences be sufficiently important to deserve a specific name. This latter point will become a far more essential consideration than it is at present; for differences, however slight, between any two forms, if not blended by intermediate gradations, are looked at by most naturalists as sufficient to raise both forms to the rank of species. Hereafter we shall be compelled to acknowledge that the only distinction between species and well-marked varieties is, that the latter are known, or believed, to be connected at the present day by intermediate gradations, whereas species were formerly thus connected. Hence, without quite rejecting the consideration of the present existence of intermediate gradations between any two forms, we shall be led to weigh more carefully and to value higher the actual amount of difference between them. It is quite possible that forms now generally acknowledged to be merely varieties may hereafter be thought worthy of specific names, as with the primrose and cowslip; and in this case scientific and common language will come into accordance. In short, we shall have to treat species in the same manner as those naturalists treat genera, who admit that genera are merely artificial combinations made for convenience. This may not be a cheering prospect; but we shall at least be freed from the vain search for the undiscovered and undiscoverable essence of the term species.

The other and more general departments of natural history will rise greatly in interest. The terms used by naturalists of affinity, relationship, community of type, paternity, morphology, adaptive characters, rudimentary and aborted organs, &c., will cease to be metaphorical, and will have a plain signification. When we no longer look at an organic being as a savage looks at a ship, as at something wholly beyond his comprehension; when we regard every production of nature as one which has had a history; when we contemplate every complex structure and instinct as the summing up of many contrivances, each useful to the possessor, nearly in the same way as when we look at any great mechanical invention as the summing up of the labour, the experience, the reason, and even the blunders of numerous workmen; when we thus view each organic being, how far more interesting, I speak from experience, will the study of natural history become!

A grand and almost untrodden field of inquiry will be opened, on the causes and laws of variation, on correlation of growth, on the effects of use and disuse, on the direct action of external conditions, and so forth. The study of domestic productions will rise immensely in value. A new variety raised by man will be a far more important and interesting subject for study than one more species added to the infinitude of already recorded species. Our classifications will come to be, as far as they can be so made, genealogies; and will then truly give what may be called the plan of creation. The rules for classifying will no doubt become simpler when we have a definite object in view. We possess no pedigrees or armorial bearings; and we have to discover and trace the many diverging lines of descent in our natural genealogies, by characters of any kind which have long been inherited. Rudimentary organs will speak infallibly with respect to the nature of long-lost structures. Species and groups of species, which are called aberrant, and which may fancifully be called living fossils, will aid us in forming a picture of the ancient forms of life. Embryology will reveal to us the structure, in some degree obscured, of the prototypes of each great class.

When we can feel assured that all the individuals of the same species, and all the closely allied species of most genera, have within a not very remote period descended from one parent, and have migrated from some one birthplace; and when we better know the many means of migration, then, by the light which geology now throws, and will continue to throw, on former changes of climate and of the level of the land, we shall surely be enabled to trace in an admirable manner the former migrations of the inhabitants of the whole world. Even at present, by comparing the differences of the inhabitants of the sea on the opposite sides of a continent, and the nature of the various inhabitants of that continent in relation to their apparent means of immigration, some light can be thrown on ancient geography.

The noble science of Geology loses glory from the extreme imperfection of the record. The crust of the earth with its embedded remains must not be looked at as a well-filled museum, but as a poor collection made at hazard and at rare intervals. The accumulation of each great fossiliferous formation will be recognised as having depended on an unusual concurrence of circumstances, and the blank intervals between the successive stages as having been of vast duration. But we shall be able to gauge with some security the duration of these intervals by a comparison of the preceding and succeeding organic forms. We must be cautious in attempting to correlate as strictly contemporaneous two formations, which include few identical species, by the general succession of their forms of life. As species are produced and exterminated by slowly acting and still existing causes, and not by miraculous acts of creation and by catastrophes; and as the most important of all causes of organic change is one which is almost independent of altered and perhaps suddenly altered physical conditions, namely, the mutual relation of organism to organism,—the improvement of one being entailing the improvement or the extermination of others; it follows, that the amount of organic change in the fossils of consecutive formations probably serves as a fair measure of the lapse of actual time. A number of species, however, keeping in a body might remain for a long period unchanged, whilst within this same period, several of these species, by migrating into new countries and coming into competition with foreign associates, might become modified; so that we must not overrate the accuracy of organic change as a measure of time. During early periods of the earth's history, when the forms of life were probably fewer and simpler, the rate of change was probably slower; and at the first dawn of life, when very few forms of the simplest structure existed, the rate of change may have been slow in an extreme degree. The whole history of the world, as at present known, although of a length quite incomprehensible by us, will hereafter be recognised as a mere fragment of time, compared with the ages which have elapsed since the first creature, the progenitor of innumerable extinct and living descendants, was created.

In the distant future I see open fields for far more important researches. Psychology will be based on a new foundation, that of the necessary acquirement of each mental power and capacity by gradation. Light will be thrown on the origin of man and his history.

Authors of the highest eminence seem to be fully satisfied with the view that each species has been independently created. To my mind it accords better with what we know of the laws impressed on matter by the Creator, that the production and extinction of the past and present inhabitants of the world should have been due to secondary causes, like those determining the birth and death of the individual. When I view all beings not as special creations, but as the lineal descendants of some few beings which lived long before the first bed of the Silurian system[24] was deposited, they seem to me to become ennobled. Judging from the past, we may safely infer that not one living species will transmit its unaltered likeness to a distant futurity. And of the species now living very few will transmit progeny of any kind to a far distant futurity; for the manner in which all organic beings are grouped, shows that the greater number of species of each genus, and all the species of many genera, have left no descendants, but have become utterly extinct. We can so far take a prophetic glance into futurity as to fortell that it will be the common and widely-spread species, belonging to the larger and dominant groups, which will ultimately prevail and procreate new and dominant species. As all the living forms of life are the lineal descendants of those which lived long before the Silurian epoch, we may feel certain that the ordinary succession by generation has never once been broken, and that no cataclysm has desolated the whole world. Hence we may look with some confidence to a secure future of equally inappreciable length. And as natural selection works solely by and for the good of each being, all corporeal and mental endowments will tend to progress towards perfection.

It is interesting to contemplate an entangled bank, clothed with many plants of many kinds, with birds singing on the bushes, with various insects flitting about, and with worms crawling through the damp earth, and to reflect that these elaborately constructed forms, so different from each other, and dependent on each other in so complex a manner, have all been produced by laws acting around us. These laws, taken in the largest sense,

24. *Silurian system*: A system of rocks containing fossils of the first fish and land plants, currently dated between 438 and 408 million years ago.

being Growth with Reproduction; Inheritance which is almost implied by reproduction; Variability from the indirect and direct action of the external conditions of life, and from use and disuse; a Ratio of Increase so high as to lead to a Struggle for Life, and as a consequence to Natural Selection, entailing Divergence of Character and the Extinction of less improved forms. Thus, from the war of nature, from famine and death, the most exalted object which we are capable of conceiving, namely, the production of the higher animals, directly follows. There is grandeur in this view of life, with its several powers, having been originally breathed into a few forms or into one; and that, whilst this planet has gone cycling on according to the fixed law of gravity, from so simple a beginning endless forms most beautiful and most wonderful have been, and are being, evolved.

Bibliography and Suggestions for Further Reading

Darwin, Charles, *The Autobiography of Charles Darwin*, many editions. Fascinating reflections on his personal life, his scientific work, and on his deepest convictions.

Darwin, Charles, *The Journal of Researches*, also called *The Voyage of the Beagle*, many editions. Darwin's own account of the voyage that shaped his scientific life. A wonderful account of science in the making.

Darwin, Charles, *The Origin of Species*. Penguin Books, 1968. Complete text of the First Edition (1859) along with the useful glossary that Darwin provided in later editions.

Bowler, Peter J., *Evolution, The History of an Idea*, University of California Press, 1989. A good general survey of evolutionary ideas.

Browne, Janet, *Charles Darwin, Voyaging*, Alfred A. Knopf, 1995.

and

Browne, Janet, *Charles Darwin, The Power of Place*, Alfred A. Knopf, 2002.

Together, these two volumes are the best biography of Darwin, comprehensive, deeply knowledgeable, beautifully written.

Depew, David J., and Bruce H. Weber, *Darwinism Evolving: System Dynamics and the Genealogy of Natural Selection*, MIT Press, 1997. A collection of essays on the development of the research tradition in evolutionary biology. Very informative.

Futuyma, Douglas, *Evolutionary Biology*, 2nd edition, Sinauer Associates Inc., 1986. A solid and reliable survey of evolutionary biology today. A good place to start to learn about the modern science. Although some parts are quite technical, much of this book can be appreciated by the general reader.

Hodge, Jonathan and Gregory Radick, eds., *The Cambridge Companion to Darwin*, Cambridge University Press, 2003. A very good collection of essays primarily concerned with philosophical questions raised by Darwin's work.

Kohn, David, *The Darwinian Heritage*, Princeton University Press, 1985. A collection of essays about many aspects of Darwin's work and its historical context. The essays are of varying quality, but some are very good, and many are informative.

Lewontin, Richard, and Richard Levins, *The Dialectical Biologist*, Harvard University Press, 1985. A collection of essays, most of which are by Lewontin, who is one of the foremost evolutionary geneticists of our time. His reflections on evolutionary biology never fail to be imaginative and interesting. Eminently readable.

Ospovat, Dov, *The Development of Darwin's Theory*, Cambridge University Press, 1981. An illuminating account of how Darwin arrived at his theory in the period 1838–1859.

Sober, Elliot, *The Nature of Selection*, University of Chicago Press, 1984. An intelligent and deep account of the concept of selection, written by a philosopher of science. Difficult and densely argued, but worth the effort.

Wilson, Edward O., *The Diversity of Life*, W. W. Norton & Co., 1992. An appreciation of the abundance of life on earth, with concerns about its fragility.

Index

About Nicholas Maistrellis

Nicholas Maistrellis teaches in the Great Books program at St. John's College, Annapolis, Maryland, where he has been especially active in shaping the laboratory curriculum. Since 1967, he and his students have been reading Darwin's works in the context of the broad western intellectual tradition. As editor of the present volume, Maistrellis draws upon a background in general biology and the history and philosophy of science.

About This Series

SCIENCE CLASSICS FOR HUMANITIES STUDIES is a series of study modules designed to bring fundamental works of science and mathematics within the grasp of students and other readers without the need for specialized preparation. The series reflects the Green Lion's conviction that an understanding of science, and especially of the classical works of science, is essential for all students of the humanities. Science, no less than poetry or philosophy, is human thought, a response both to the outer world of our senses and the inner experience of our consciousness. The more profound a scientific work is, the more directly it addresses itself to our humanity; therefore, there is much in the greatest works of science that can be grasped without special preparation. Yet too many educational programs find themselves limited by the supposed divide between the humanities and the sciences—the so-called "two cultures."

Further, teachers and institutions who wish to heal this unnecessary fracture have had to confront two discouraging barriers. On the one hand, classic texts of real science are often found to be forbiddingly technical in content and burdened with terminologies either antiquated or arcane. On the other hand, popularizations of these classics insulate students from the actual workings of thought and imagination that classic texts embody. Green Lion Press has addressed this dilemma with the series SCIENCE CLASSICS FOR HUMANITIES STUDIES, issued in slim, inexpensive student editions under the *Green Cat Books* imprint.

Each volume in the series is a compact, inexpensive presentation of classic scientific and mathematical texts, offering generous but judicious guidance for the reader. We have drawn on our many years of reading these books with nonspecialist students to choose selections of real substance, and to provide helps that make the texts accessible while at the same time allowing the original authors to speak for themselves, in their own voices.

Besides humanities students, this series will appeal to those interested in science but lacking time or expertise to read these works unabridged and without assistance. It will also serve readers who already enjoy a technical background but who may wish to experience more directly the sources and often surprising origins of contemporary scientific concepts.

Other Books In This Series

Selections from Newton's *Principia*, by Dana Densmore

A compact counterpart to Densmore's acclaimed *Newton's Principia: the Central Argument,* this Module includes Newton's prefatory material, the Definitions and Scholium, the Laws of Motion with its scholium, and the concluding General Scholium. For the more adventurous readers, the book also presents Newton's proof, using only elementary geometry, that the moon is held in its orbit about the earth by the force of gravity. All is presented with extensive guidance, in the form of a general introduction, a glossary of Newton's terminology, notes on the individual sections and proofs, and questions for discussion.

Faraday's *Experimental Researches in Electricity:* The First Series, by Howard J. Fisher

In this volume, containing Michael Faraday's first published electrical investigations, Faraday boldly explores a new relation between magnetism and electricity. As we follow his narrative, aided by Howard Fisher's introductions, notes, and supplementary diagrams, we can both witness and participate in his thinking, his questioning, and his insights.

Selections from Kepler's *Astronomia Nova,* by William H. Donahue

Johannes Kepler wrote *Astronomia Nova* (1609) in a concerted drive to sweep away the ancient and medieval clutter of spheres and orbs and to establish a new truth in astronomy, based on physical causality. Accordingly, much of the book is given over to a nontechnical discussion of how planets can be made to move through space by physical forces. This is the theme of the readings in the present module. Also included are Kepler's thoughts on the nature of gravity, the relation between scriptural and scientific truth, and the development of Kepler's first two laws of planetary motion.

Euclid's *Elements,* Book One with Questions for Discussion, by Dana Densmore

Book One of the *Elements* is a beautiful and coherent composition, moving in sometimes surprising ways to a very satisfying end. It is deeply rewarding to readers who are ready to go beyond merely following step by step, who are interested in *why* this step here, *how* learning and discovery happen, and what a particular proof shows us about what it means for things to be the same or different. Rather than serving as commentary or explanation of the text, Dana Densmore's accompanying Questions for Discussion are intended as examples, to urge students to think more deeply and carefully about what they are watching unfold, and to help them find their own questions in a genuine and exhilarating inquiry.